Questions and Answers in Electrical Installation Technology

By the same author

Electrical Installation Technology 1
Electrical Installation Technology 2
Electrical Installation Technology 3

Questions and Answers in Electrical Installation Technology

Maurice Lewis

B. Ed(Hons), FIElecIE

Hutchinson Education

London Sydney Auckland Johannesburg

Text © Maurice Lewis 1989

All rights reserved. No part of this publication may be reproduced or transmitted in any form or by any means, electronic or mechanical, including photocopy, recording, or any information storage and retrieval system, without permission in writing from the publisher or under licence from the Copyright Licensing Agency Limited. Further details of such licences (for reprographic reproduction) may be obtained from the Copyright Licensing Agency Limited, of 90 Tottenham Court Road, London W1P 9HE.

First published in 1989 by Hutchinson Education

Reprinted in 1991 by
Stanley Thornes (Publishers) Ltd
Old Station Drive
Leckhampton
CHELTENHAM GL53 0DN
England

British Library Cataloguing in Publication Data

Lewis, M. L. (Maurice L)
    Questions and answers in electrical installation technology
    1. Electric equipment. Installation
    I. Title
    621.31'042

ISBN 0 7487 0236 9

Typeset by Wyvern Typesetting Ltd

Printed and bound in Great Britain at
The Bath Press, Avon

Contents

Preface   7

Part I Studies
Theory and Regulations   11
Science and Calculations   37

Part II Studies
Theory and Regulations   59
Science and Calculations   89

Multiple-choice questions
Part I Certificate   123
Part II Certificate   136
Revision exercise one   150
Revision exercise two   162
Revision exercise three   174

Answers and comments to multiple-choice questions   183

Appendices
Appendix 1   Examination hints   199
Appendix 2   Laboratory work   201
Appendix 3   Example of Part I assignment answer   212
Appendix 4   Possible course study route at a technical college   218
Appendix 5   Common deviations in electrical installations   219

Preface

This book has been written for students studying City and Guilds Course 236 Part I and Part II certificates in electrical installation work. There are 200 worked examples, many of which are illustrated, and the chosen topics make it a suitable textbook for other electrical students, particularly those following Course 201 and Course 232.

A number of the worked examples can be found as questions in volume 1 and volume 2 of Electrical Installation Technology by the same author. Also included are numerous City and Guilds exam questions from recent papers.

In order to provide a spread of the syllabus, the book includes some 500 multiple-choice questions and students will not only find the answers to these questions in the book, but also find helpful suggestions and hints to aid their learning process. A wide selection of topics are in keeping with the National Council for Vocational Qualification practical competences.

In view of City and Guilds examination reports continually mentioning candidates' poor layout, poor description and poor diagrams, etc., it has been decided to include an appendix on examination hints, one with a model answer for an assignment, and a further appendix on laboratory work. The latter includes several useful experiments in order to show students how to express their findings on paper and to try to improve this area of written communication. For clarity, the author has referred experiment calculations to coursework notes but it should be noted that college lecturers might insist that all calculations and workings are properly shown in either the results section of an experiment or in a separate appendix at the end.

Grateful acknowledgements are made to the Institution of Electrical Engineers for allowing the IEE Regulations to be quoted and also to the City and Guilds of London Institute for allowing past paper questions to be used. All solutions are those of the author.

M.L.L.
1989

Part I Studies

Theory and Regulations

1. What precautions should be taken when:
   (a) handling heavy equipment
   (b) carrying a ladder and then erecting it in position
   (c) working with oily or greasy materials
   (d) drilling holes with a portable electric drill
   (e) terminating electrical conductors inside switchgear

   Solution

   Some precautions are as follows:
   (a) Try to seek assistance.
   Use appropriate lifting tackle.
   Wear appropriate clothing – protective gloves and shoes.
   Be aware of the people around you and take care not to damage things in transit.
   (b) Carry ladder with front elevated.
   Stand ladder on firm base and use 4:1 rule for climbing.
   Lash ladder at top or secure the bottom.
   Only one person should be on a ladder and it is dangerous to over-reach.
   (c) Use barrier cream on hands.
   Be aware of the material slipping from grasp.
   Wipe tools clean of oil or grease before use.
   (d) Do not exert additional pressure on drill.
   Keep lead away from the drilling area.
   Be aware of the drill bit as it pierces the hole, scattering hot metal fragments.
   Keep drill straight.
   Clean up area where drilling took place.
   (e) Allow sufficient tails for terminating.
   Avoid stress on conductors.
   Check insulation for damage.
   Tighten terminals and label conductors.
   Check number of tools used.
   Fasten enclosures.

2. Briefly state **two** legislative requirements of the Health and Safety at Work Act 1974 for:
   (a) an employer
   (b) an employee

   Solution

   (a) It is the duty of every employer to ensure, so far as reasonably practicable, the health, safety and welfare at work of all his employees.
   An employer must also provide for his employees, so far as reasonably practicable, information, instruction, training and supervision.
   (b) It is the duty of every employee to exercise reasonable care for the health and safety of himself and others who may be affected by his acts or omissions at work.

An employee must also co-operate with his employer, so far as may be necessary, to enable him to carry out his legal duties regarding health and safety matters.

3 Comment on the legal requirements with regard to:
(a) eye protection
(b) protection of skin

Solution

(a) The employer has a legal duty under Section 2(2) of HSW Act to provide eye protection for his employees and to ensure that it is worn where there is any forseeable risk. Types of hazard might include impact from solids, ingress from liquids, dust or gas, and even exposure to glare. Types of eye protector have to be carefully chosen and meet the recognised British Standards.
(b) Employers have the responsibility under the HSW Act Section 2(2) for ensuring safe systems of work are used and adequate welfare facilities are available. They also have the responsibility of providing protective clothing, first aid and washing facilities. Barrier and cleansing creams should be available. Pitch, bitumen, cement, lime, petrol, fibre glass, acids and alkalis are some examples of harmful substances.

4 Make a sketch of a 13 A plug top and fully lable its internal connections and parts. Describe how you would terminate a flex into the plug top.

Solution

A sketch of the plug top is shown in Figure 1 with the essential parts labelled. Essential points to note are:
(a) Only strip sufficient insulation away from internal conductors.
(b) Anchor strands on terminals in a

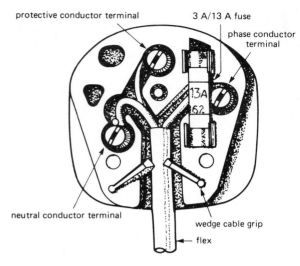

Figure 1 *13 A plug top*

clockwise direction and do not screw terminal over the insulation. Check polarity.
(c) Insert flex in the wedge cable grip before fitting plug cover.
(d) Remember to check the fuse size for the equipment being used.

5 A two-way lighting circuit has become faulty due to a recent alteration in the switch positioning. Describe the precautions taken in rectifying the fault and mention the possible troubles with the circuit wiring.

Solution

If only one switch has been altered then it is quite obvious that the trouble will be here. However, keep an open mind in case it so happens that a fault has developed elsewhere. First, switch off the circuit supply or remove the lighting fuse, whichever is convenient. Since the question does not indicate what kind of fault exists let us assume it is one where the light sometimes doesn't work from one switching position and only works in one direction from the other switching position. The

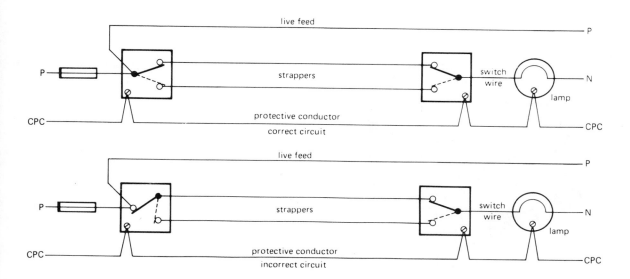

Figure 2 *Two-way switching*

cause of the fault here will be that the live feed has been inserted in one of the strapper wire terminals. If the light does not work at all, it is possible that the conductor feed or switch wire has broken off inside its terminal. If the fault cannot be traced visually by inspection at the altered switch position and the fault is one of no light, check the lamp itself and the fuse – although the fuse fault would have meant other circuits not working. Where the fault causes the fuse to rupture, then it is likely to be an earth fault in the switch. Check wiring polarity and earth sleeving. See Figure 2 for possible cause.

6   Why is it sometimes necessary to earth the metal cover of a 13 A socket outlet to its box? What size protective conductor is normally used for this purpose?

Solution

In practice, where metal conduit is used to provide the protective conductor of a wiring system, then socket outlet boxes attached to the conduit must provide good earth continuity. The metal cover plate of these boxes is held fast by two fixing screws which are inadequate as a means of providing continuity with the rest of the box, particularly if the mounting box is for flush wiring having an adjustable lug (see Figure 3). It is a requirement of the IEE Wiring Regulations, Reg. 543–10 that a separate protective conductor must connect the box to the socket outlet via an earthing terminal incorporated within the box. The minimum size protective conductor is 2.5 mm^2. See IEE Wiring Regulations, Reg. 547–4.

7   Explain the procedure you would take in attempting to rescue a fellow workmate who was receiving an electric shock whilst holding a portable drill.

Solution

Some guidance notes on electric shock treatment are to be found on pages 21–22, *Electrical Installation Technology 1*. One's first reaction is to immediately switch off the electrical supply. Since the tool is a portable drill its lead will not be very long and should easily be traced to a supply source. Where this

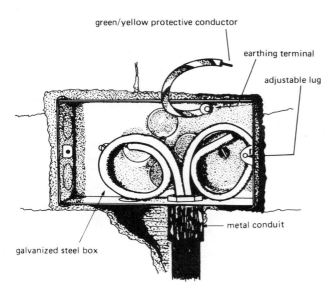

Figure 3 *Earth connection*

may not be the case, your workmate has to be released from contact with the drill using some form of non-conducting material, such as rubber gloves, dry clothing, dry wood or a length of PVC tubing.

If your workmate is in an unconscious state and not breathing (looking pale or even blue if the airway is blocked), start mouth-to-mouth first aid treatment. Only a few seconds delay can mean the difference between success and failure. The procedure is given below supported by Figure 4.

(a) Kneel by your workmate's head and quickly inspect the mouth for any obstructions. Loosen clothing around the neck.

(b) Move the head fully back (as shown), breathe in deeply, seal your lips over his or hers, pinch the nostrils with one hand and breathe out into the body.

(c) Watch the chest rise and then turn your head away and breathe in once more. Repeat the procedure about ten to twelve times every minute and continue until he or she is breathing satisfactorily or until you are told by a doctor to stop.

(d) If your workmate recovers before the doctor arrives (someone should have been sent to get help), keep him or her warm and place in a recovery position.

A further stage in the treatment of electric shock is described in *Electrical Installation Technology 1*, but carrying out this method (external heart compression) requires first-aid training.

8 Figure 5 shows the main details required on an accident form. Complete the blank form for an accident involving the treatment of electric shock as described in Question 7.

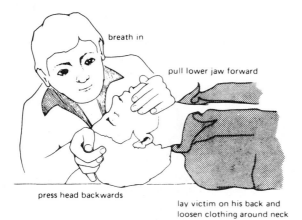
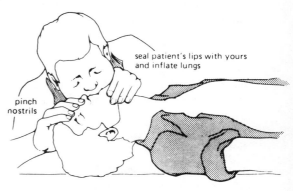

Figure 4 *Mouth-to-mouth ventilation*

NOTICE OF ACCIDENT OR DANGEROUS OCCURRENCE

1. OCCUPIER OF PREMISES Name Address Nature of Business	4. PLACE WHERE INCIDENT OCCURRED Address Exact Location (e.g. staircase to office, canteen, storeroom, classroom).
2. EMPLOYER OF INJURED PERSON (if different from above) Name Address	Name of Person supervising
	5. INJURIES AND DISABLEMENT Fatal or non-fatal Nature and extent of injury (e.g. fracture of leg, laceration of arm, scalded foot, scratch on hand followed by sepsis).
3. INJURED PERSON Resident/Staff Surname Widow/Widower Christian Names Married/Single Date of Birth Occupation Address Name and address of parent or guardian	

6. ACCIDENT OR DANGEROUS OCCURRENCE
Date   Time
Full details of how the incident occurred and what the injured person was doing. If a fall of person or materials, plant, etc. state height of fall.
Name and address of any witness.
If due to machinery, state name and type of machine. What part of the machine caused the accident? Was the machine in motion by mechanical power at the time?

7. ACTION FOLLOWING THE ACCIDENT
What happened?
When was the doctor informed?   Name of Doctor When did he attend?   (address and telephone) If taken to hospital, say when and where.
Names and addresses of friends or relatives who have been notified of the accident:
When and how were they informed?
Signature of injured person or person completing this form: *Date:* *If the form is completed by some person acting on the injured person's behalf, the address and occupation of such person should be entered.*

Figure 5 *Notice of accident or dangerous occurrence*

The solution to question 8 appears on page 16.

Solution

NOTICE OF ACCIDENT OR DANGEROUS OCCURRENCE

1. OCCUPIER OF PREMISES Name **J.A. TOWNSHOTT** Address **5, APPLEBY WAY, ELY** Nature of Business **BRICKWORKS**	4. PLACE WHERE INCIDENT OCCURRED Address **5 APPLEBY WAY ELY** Exact Location (e.g. staircase to office, canteen, storeroom, classroom). **STOREROOM** Name of Person supervising **J. FLASH**
2. EMPLOYER OF INJURED PERSON (if different from above) Name Address	5. INJURIES AND DISABLEMENT Fatal or non-fatal **NON FATAL** Nature and extent of injury (e.g. fracture of leg, laceration of arm, scalded foot, scratch on hand followed by sepsis). **ELECTRIC SHOCK FROM FAULTY PORTABLE DRILL**
3. INJURED PERSON ~~Resident~~/Staff ~~Widow/Widower~~ Married/~~Single~~ Surname **SPARK** Christian Names **IAN** Date of Birth **6.7.38** Occupation **ELECTRICIAN** Address **12 OAKLANE WOOTEN** Name and address of parent or guardian **Mr. & Mrs. J.A. SPARK** **108 ELM DRIVE, WOOTEN**	

6. ACCIDENT OR DANGEROUS OCCURRENCE
Date **13TH AUGUST 1988**   Time **11-00 am**
Full details of how the incident occurred and what the injured person was doing.
If a fall of person or materials, plant, etc. state height of fall.

Mr. SPARK WAS DRILLING A METAL BOX AT THE TIME OF THE ACCIDENT

Name and address of any witness.
Mr. P. BROWN
22 WINDSOR ROAD, LILLY, KENT

If due to machinery, state name and type of machine.
What part of the machine caused the accident?
Was the machine in motion by mechanical power at the time?

7. ACTION FOLLOWING THE ACCIDENT
What happened? **Mr. SPARK WAS TREATED FOR ELECTRIC SHOCK**
When was the doctor informed? **11·00 am**
When did he attend? **11·15 am**

Name of Doctor **P. S. BONE** (address and telephone)
BROMHAM LANE SURGERY
BROM 61062

If taken to hospital, say when and where.
11-45 am   UPTON HOSPITAL
Names and addresses of friends or relatives who have been notified of the accident:
PARENTS (SEE ABOVE)
When and how were they informed? **11-45 am - TELEPHONE**

Signature of injured person or person completing this form:
J. Flash   Date: **13-8-83**
If the form is completed by some person acting on the injured person's behalf, the address and occupation of such person should be entered.

9 Explain the use of the following circuit components:
(a) power factor correction capacitor
(b) choke
(c) current transformer
(d) voltage transformer
(e) contactor

Solution

(a) A *power factor correction capacitor* is a component which neutralizes the effect created in an inductive circuit, such as by a choke, transformer winding or motor field winding, when the components are connected to an a.c. supply. These components tend to have lagging power factors whereas the capacitor component has a leading power factor.

(b) A *choke* is an inductive component used to operate a discharge lamp; it may sometimes be called a *ballast*. In a fluorescent tube circuit, the choke provides the initial voltage surge to strike an arc across the lamp electrodes. Once this occurs the choke then protects the circuit by limiting the amount of current through the lamp.

(c) A *current transformer* is an instrument transformer used for measurement whereby, for example, an ammeter, usually designed to give a full scale deflection of 5 A, can in fact be used with the transformer to read a small proportion of a large load current. The current transformer allows the use of much smaller cross-sectional area conductors for measurement purposes.

(d) A *voltage transformer* is again an instrument transformer used for measuring low voltage from high voltage installations; they make it possible to use standard low voltage cables for measurement since their measuring voltage is usually at 110 V.

(e) A *contactor* is a power control device comprising a magnetic core, operating coil and associated contacts. It is used for the automatic opening and closing of circuits.

10 With the aid of circuit diagrams show the correct connections for the following meters in a single-phase a.c. supply system:
(a) kilowatt-hour energy meter
(b) power factor meter
(c) wattmeter
(d) ammeter
(e) voltmeter

Solution

See Figure 6.

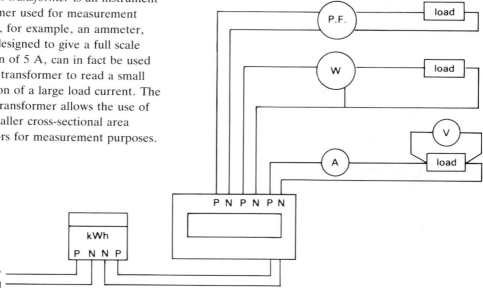

Figure 6 *Energy meter, power factor meter, wattmeter, ammeter and voltmeter connections*

11 (a) State **one** example of each of the heating, magnetic and chemical effects of electric current.
(b) Briefly explain with the aid of a sketch an example of the heating effects of current.

Solution

(a) A storage radiator is an example of the heating effects of electric current.
An electric motor is an example of the magnetic effects of electric current.
A secondary cell is an example of the chemical effects of electric current.

(b) Figure 7 shows the essential components of a storage radiator. It is able to retain its heat because of the very good thermal conductivity of its refractory bricks and thermal insulation – the latter being mineral wool or glass fibre lining inside the heater's casing. The heater is normally on during off-peak hours between 1.0 and 7.0 a.m. using an Economy 7 tariff.

12 Briefly explain the following terms and state their units:
(a) impedance
(b) reactance
(c) resistance

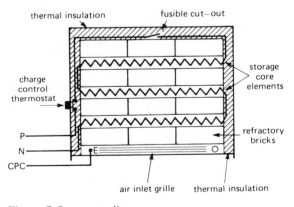

Figure 7 *Storage radiator*

Solution

(a) *Impedance* in an a.c. circuit can be regarded as the total opposition to current flow combining together both resistance and reactance in the circuit.
(b) *Reactance* is the property of an inductor or capacitor.
(c) *Resistance* is the property of a resistive component as well as being the property of an inductive coil.
All three terms are expressed in ohms.

13 In the diagram of Figure 6, briefly explain the instrument connections

Solution

(a) The kilowatt-hour energy meter is an integrating meter which records energy (power × time). It is connected at the intake position.
(b) The power factor meter measures three conditions of power factor, namely unity p.f., lagging p.f. or leading p.f. It requires four connections.
(c) The wattmeter measures power (volts × amps) consumed in a circuit and requires connections for its voltage coil and current coil.
(d) The ammeter measures current and is connected in series with the load.
(e) The voltmeter measures electrical pressure or potential difference and is connected across terminals of differing potentials such as the supply terminals or load terminals.

Note: Power factor can be determined by using a wattmeter, ammeter and voltmeter. A circuit for this will be found later in the book.

14 Explain the principles of operation of simple a.c. and d.c. generators.

Solution

The principles are based on electromagnetic

induction. When a conductor is rotated within a magnetic field, the invisible magnetic flux induces a voltage in the conductor. An induced current will flow if the conductor forms a closed circuit. A.C. generators derive their output through slip-rings whereas d.c. generators derive their output through a device called a commutator. For further understanding, see *Science and Calculations* by the same author.

15 Write down meanings for the following:
(a) open circuit
(b) closed circuit
(c) short circuit
(d) polarity
(e) continuity
(f) circuit breaker
(g) insulation resistance
(h) fusing factor
(i) current rating
(j) space factor

Solution

(a) *open circuit* This generally refers to an electrical circuit that has become broken or discontinuous so current cannot flow.
(b) *closed circuit* This generally refers to an electrical circuit that is continuous so that current will flow.
(c) *short circuit* This generally refers to a fault condition whereby live conductors are shorted out.
(d) *polarity* In electrical terms this is an indication of conductor polarity whether it is a 'phase', 'neutral' or 'earth' conductor or whether it is of 'positive' or 'negative' polarity.
(e) *continuity* In electrical terms, this implies a cable's ability to conduct a continuous flow of current through circuit conductors. Earth continuity implies an uninterrupted path, preferably of low resistance.
(f) *circuit breaker* A mechanical/electrical device designed to open or close a circuit under normal or abnormal conditions. A miniature moulded case circuit breaker is an example.
(g) *insulation resistance* The insulation medium surrounding and supporting a live or potentially live conductor. Its ohmic value should be very high.
(h) *fusing factor* This is a factor which expresses the minimum fusing current at which a fuse element will melt divided by the current rating.
(i) *current rating* The maximum current that a fuse will carry without exceeding a specified temperature rise.
(j) *space factor* This is a ratio of the sum of the overall cross-sectional areas of cables (including sheath) to the internal cross-sectional area of their enclosure such as conduit and trunking.

16 What is meant by the two terms: 'overcurrent' and 'fuse discrimination'?

Solution

Overcurrent is a circuit condition which results in either an overload or a dead short (see short circuit above). The former occurs in a perfectly sound circuit which is not regarded as an electrical fault. The latter occurs as a result of an electrical fault between live conductors or live conductor(s) and earth. Fuse discrimination is the ability of an overcurrent protective device to interrupt the supply to a circuit when a fault has developed without affecting healthy circuits in the same system. It is generally assumed that the fuse (minor fuse) nearest the fault downstream from the origin of the circuit should rupture first.

17 Explain the operation of a residual current device.

Solution

The device is described on page 48 of *Electrical Installation Technology 1* and illustrated in Figure 8. It will be seen that both live conductors, i.e. phase and neutral conductors, pass around a toroidal transformer core before connection is made to the load circuit. Both windings have an equal and opposite number of turns so that no flux will be induced in the core to set up a circulating trip current through the detector coil circuit. This will only be the case when the device detects no earth fault. A short circuit between phase and neutral would also not cause the device to trip out since its operation has to detect an imbalance in the phase and neutral currents on each side of the core. Any leakage to earth on the load side in one of the live conductors allows the ampere-turns (induced flux) to be greater in the other conductor and circulate a flux around the core. Notice that the detector coil and trip coil form a closed circuit.

18 Why is it good practice to balance phases in a.c. three-phase, four-wire systems?

Solution

The reason for this practice is to establish proper selection of main switchgear and main cables for the wiring system. If one phase of the supply was over-assessed while the other phases were under-assessed, then selection of main switch and cables would have to be based on the over-assessed phase. If the system was balanced, i.e. all phases designed to carry equal current, then smaller switchgear and cables could be selected. This would lead to a more efficient and economical installation.

19 Draw a 415 V/240 V distribution system and indicate on the drawing both line and phase voltages between red, yellow and blue lines and between each phase and neutral. Show also how 64 V can be obtained through a double-wound transformer.

Solution

See Figure 9.

20 A domestic consumer's previous and present quarterly meter readings are 58 142 and 59 361 respectively. If each unit of electricity costs 4.83p and the standing charge is £6.37, what is his quarterly bill?

Solution

It will be noticed that there are two charges being made, the first is a unit charge, the second a standing or fixed charge.

The number of units used is the difference between both sets of reading, i.e.:

$$\begin{array}{r} 59\ 361 \\ -58\ 142 \\ \hline 1\ 219 \end{array}$$

At the cost of 4.83p/unit

	£
The unit charge is	58.88
Add standing charge	6.37
Total cost	65.25

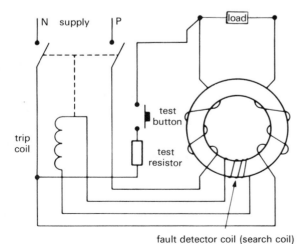

Figure 8 *Residual current device*

Theory and Regulations 21

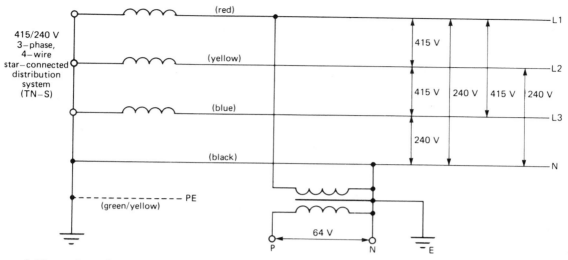

Figure 9 *Three-phase, four-wire distribution system*

21 Describe with the aid of a diagram the operation of a fluorescent lamp and its associated control-gear.

Solution

The circuit is shown in Figure 10 and its operation is explained on page 82–83 of *Electrical Installation Technology 1*.

It will be seen from the diagram that the fluorescent tube circuit consists of a choke or lamp ballast, power factor correction capacitor and starter switch. The lamp ballast is in series with the fluorescent tube while the starter switch is connected across the tube electrodes and the correction capacitor across the supply terminals of the lamp circuit.

When the circuit is connected to the supply, a p.d. occurs across the starter switch contacts which are bimetallic strip electrodes enclosed in a neon filled glass tube. The p.d. causes the neon to glow and heat up the electrodes so that they come together and allow the passage of current around the circuit. The current heats up both lamp electrodes but because no p.d. exists in the starter switch at this moment, its

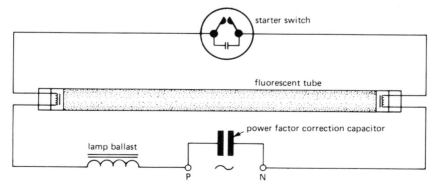

Figure 10 *Fluorescent lamp circuit*

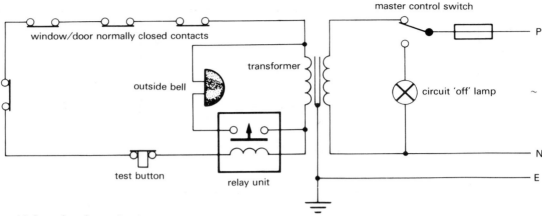

Figure 11 *Intruder alarm circuit*

bimetallic contacts cool down and spring apart. The breaking of these contacts causes the choke to produce a momentary high voltage that is sufficient to strike an arc in the fluorescent tube and its internal phosphor coating converts ultra-violet radiation into visible light.

The starter switch is provided with a radio interference suppressor because of its contacts opening and closing. The power factor correction capacitor is fitted because of the choke's poor power factor, making the circuit take more current than necessary.

22 Draw a simple diagram of a closed-circuit intruder alarm circuit consisting of four contact points, transformer, relay, indicator lamp and external bell.

Solution

See Figure 11.

23 Distinguish between:
(a) *emergency lighting* and *escape lighting*
(b) *maintained emergency lighting* and *non-maintained emergency lighting*.

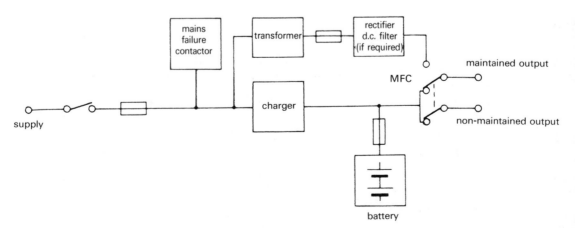

Figure 12 *Schematic diagram of emergency lighting system*

Theory and Regulations

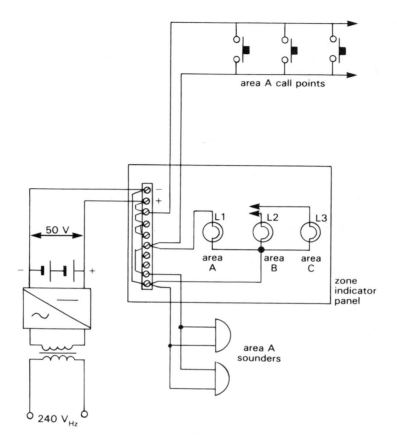

Figure 13 *Simple fire alarm system*

Solution

(a) Emergency lighting is a general term to denote lighting intended to allow occupants of a building to see in the event of normal lighting failure. Escape lighting by its implication is emergency lighting which ensures that the means of escape can be safely and effectively used at all material times.

(b) A maintained emergency lighting system is one in which all emergency lamps are in operation at all material times, whereas a non-maintained emergency lighting system is one in which all emergency lamps are in operation only when the normal lighting fails.

Note: Figure 12 shows a typical line diagram of a maintained and non-maintained output for emergency supplies.

24 Draw a circuit diagram of a simple open circuit fire alarm system operated with three call points, two sounders and a zone indicator panel. Show the method of transforming and converting the 240 V a.c. supply into 50 V d.c.

Solution

See Figure 13.

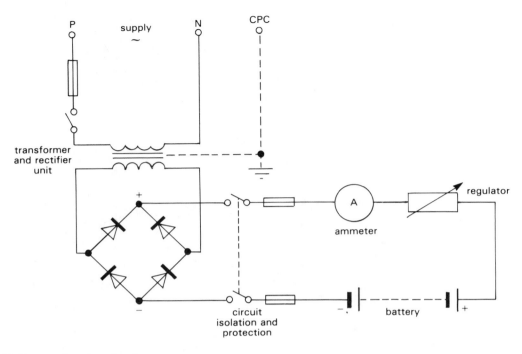

Figure 14 *Battery charging circuit*

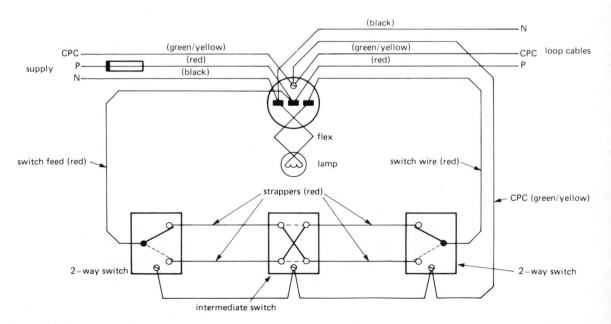

Figure 15 *Two-way and intermediate switching*

25 Make a neat circuit diagram of a simple battery charger, comprising double-wound transformer, bridge rectifier, ammeter, regulator and battery being charged.

Solution

See Figure 14.

26 Draw the single-pole switching arrangements for two-way and intermediate control of a lamp. Label all conductors in the circuit.

Solution

See Figure 15.

27 Draw a circuit diagram of 13 A radial circuit feeding six socket outlets. If the circuit covers an area greater than 30 m² and it is wired in MIMS cable, indicate on the drawing the size of the conductors used and the final circuit fuse.

Solution

See Figure 16.

28 Draw the ring final circuit of Figure 102 in *Electrical Installation Technology 1* and incorporate 13 A socket outlets at:
(a) the origin of the circuit
(b) a socket outlet on the ring
(c) a joint box half-way round the ring

Solution

See Figure 17.

29 Where on a ring circuit is it possible to wire and connect non-fused spurs? How many socket outlets (13 A) can be fed by non-fused spurs?

Solution

Reference should be made to the IEE Wiring Regulations, page 114, Appendix 5. Briefly, from the point of view of BS 1363 socket outlets, a non-fused spur feeds only one single or one twin socket outlet or one permanently connected piece of equipment and must be connected at a socket outlet on the ring or at a

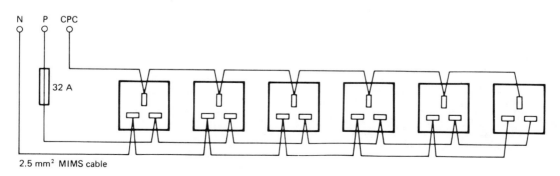

Figure 16 *Radial final circuit feeding 13-A socket outlets*

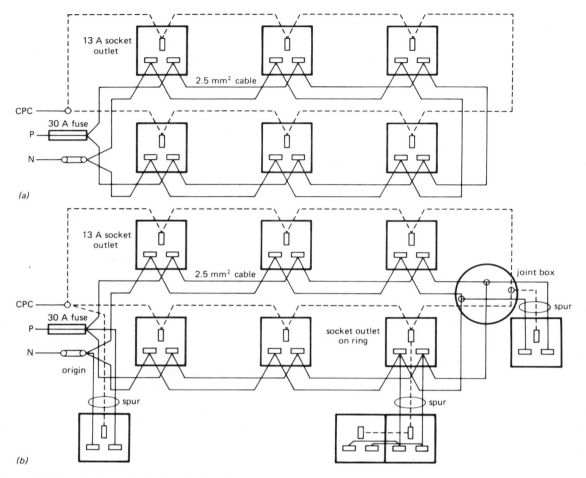

Figure 17 *Ring final circuit feeding 13-A socket outlets*

joint box, or it may be connected from the fuse position in the distribution board which supplies the ring circuit – this is called the origin of the circuit.

The number of non-fused spurs must not exceed the total number of socket outlets and stationary equipment connected within the ring circuit.

In final circuits using BS 196 socket outlets, non-fused spurs are not used.

30 Determine the assessed current demand of a 12 kW/240 V cooking appliance used in a domestic dwelling.

Solution

Reference should be made to Table 4A, Appendix 4, IEE Wiring Regulations. Here it will be seen that there is a diversity allowed. Thus:

First appliance

$$I = \frac{P}{V} = \frac{12\,000}{240} = 50 \text{ A}$$

Therefore

$$I_b = 10 + (0.3 \times R) + 5$$
$$= 10 + 12 + 5 = \mathbf{27A}$$

The question assumes the control point has facilities for a 13 A socket outlet.

31 Explain how an immersion heater thermostat operates.

Solution

A thermostat is a temperature-sensing control device that operates by cycling during normal use and which is designed to keep the temperature of an appliance between certain values. Immersion heater thermostats are generally designed to the specifications of BS 3955 and the tests prescribed therein to see if they have insufficient self-heat to cycle when their temperature-sensing elements are maintained at any suitable constant temperature.

The type often used for controlling temperature of domestic water heaters is made of brass and operates a single-pole micro-gap switch. Figure 18 illustrates this type of thermostat, where it will be seen that the brass tube and invar steel rod (non-expanding rod)

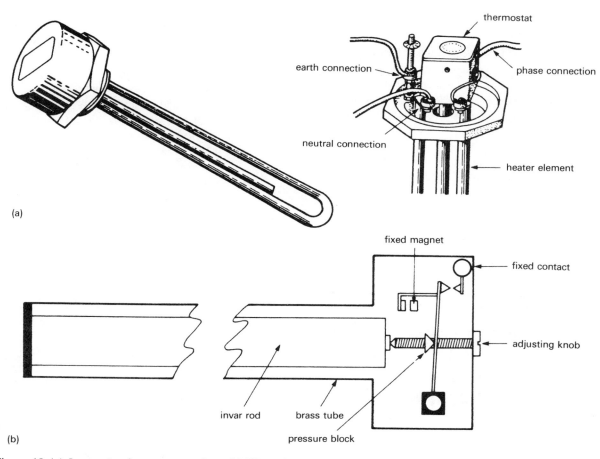

Figure 18 *(a) Immersion heater connections (b) Water heater thermostat*

are joined at one end. This is done so that when the temperature rises in the heated water, the expansion of the brass tube reduces the pressure on the pressure block and the contacts separate. The small magnet allows the mechanism to have a snap action effect to avoid unnecessary sparking and radio interference. The temperature scale and adjusting knob are fitted in the head of a moulded plastic cover. It should be pointed out that in hard water areas, the thermostat setting should not be higher than 60–65 °C.

32 With the aid of a diagram explain the operation of an oven hotplate simmerstat.

Solution

The simmerstat controller is basically a variable heat switch or energy regulator and is shown in Figure 19. It will be seen that the supply phase connection is connected to a fixed contact while a moving contact conveys current through the heating element and also through a compensating bimetal and heater winding. A control cam adjusts the initial temperature required and its contact with the bimetallic strip only allows the lower leaf of the strip to move away from the fixed contact. This occurs when the heater winding bends the bimetallic strip apart.

33 Briefly state the IEE Wiring Regulations with regard to:
(a) bathrooms and shower rooms
(b) cooking appliances

Solution

(a) Regulations appropriate to these rooms include Regs 411–2 to 411–10, SELV circuits; Reg. 413–7, earth supplementary bonding; Regs 471–34 to 471–39, which generally concern no provision for socket outlets and for the connection of portable equipment – except SELV circuits [see Reg. 471–39(a) and 471–39(b)]. Lampholders must comply with Reg. 471–38 and be shrouded with an insulating material if they are within 2.5 m of a bath or shower cubicle. Reg. 471–34 refers a maximum distance of 2.5 m for siting socket outlets in rooms (not bathrooms) containing a shower cubicle. Reg. 471–37 allows the use of BS 3052 shaver unit and earthing terminal must be connected to the final circuit c.p.c. Also, insulated cords of cord-operated switches are allowed under Reg. 471–39.

(b) One important regulation appropriate to cooking appliances is Reg. 476–20 which requires every stationary appliance to be controlled by a switch sited not more than 2 m from the appliance. If two stationary appliances are installed, then one switch can control both appliances provided that neither appliance exceeds the stated distance.
    Tables 4A and 4B of Appendix 4, IEE Regs, provide a guide to the assumed current demand of cooking appliances.

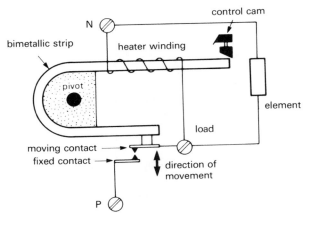

Figure 19

34 Determine the assumed current demand on a household cooking appliance rated at 12 kW/240 V if its control panel provides facility for a 13 A socket outlet.

Solution

With reference to Table 4, Appendix 4, IEE Regs, the assessment is based on a diversity allowance of first 10 A, plus 30% remainder current, plus 5 A if socket outlet is incorporated. Thus:

$$I = P/V = 12{,}000/240 = 50 \text{ A}$$
Assessment is
$$I = 10 + (0.3 \times 40) + 5 = 27 \text{ A}$$

35 Name **five** good conductors of electricity and **five** good insulators of electricity. For each, state some application.

Solution

Good conductors:
1. Copper – used as either a bare conductor (e.g. copper busbar) or an insulated conductor as a core in a cable.
2. Brass – used as a termination post to anchor electrical conductors or as switch contact points.
3. Lead – used as a terminal post on batteries or protective sheathing for cables.
4. Silicon steel – used as the core construction of transformers and the frame of motors.
5. Nichrome – used as a heating element.

Good insulators:
1. Polyvinyl chloride (PVC) – used as a cable sheath or insulated protective covering for electrical equipment.
2. Rubber – used as a cable sheath or covering and also to provide flexibility of movement.
3. Porcelain – used as an insulating medium between an earthed surface and a 'live' surface such as a spacer supporting a 'live' busbar.
4. Mica – used as an insulating medium between copper segments found on a motor's commutator.
5. Bakelite – used as an insulating medium for electrical accessories.

36 Briefly explain the essential points in the soldering of metals.

Solution

Soldering is basically the joining of metals using a low melting substance filler such as an alloy of lead and tin. In the process of soldering, a chemical reaction takes place where a compound bond is formed between the solder and metal joints. To form this compound the surfaces must be clean of surface oxides and a flux is used for this purpose. A good soldered joint requires (i) sufficient heat, (ii) a good joint design, (iii) adequate flux and (iv) a suitable solder. Some fluxes are corrosive such as 'killed spirit' and the joint should be washed after soldering. A common non-corrosive flux is called 'resin', which is used extensively for electrical work.

37 Briefly explain the following terms regarding the properties of a material:
(a) strength
(b) elasticity
(c) malleability
(d) ductility
(e) toughness and hardness

Solution

(a) The strength of a material is its ability to resist fracturing or breaking under some applied force, e.g. chain links under tension.
(b) The elasticity of a material is its ability to return to its original shape when a force or load has been removed from it, e.g. a steel spring after compression or extension. It should be noted that if the force is too great, the material becomes deformed, exhibiting 'plastic' behaviour.
(c) The malleability of a material is its ability to be rolled or hammered. Lead is more

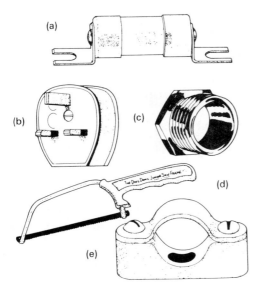

Figure 20

malleable than copper and copper is more malleable than mild steel. A blacksmith works with red hot metal because the hotter it becomes the more malleable it will be.

(d) The ductility of a material is its ability to be stretched or bent without it breaking. Materials having this property can be reshaped or easily formed by using a tool. Electrical conductors such as copper, aluminium and brass need to be ductile as well as strong.

(e) A material's toughness is its ability to withstand knocks without breaking, whereas a material's hardness is its ability to withstand scratching and wear.

38 Draw *freehand* the following objects:
(a) HBC fuse
(b) 13 A plug top
(c) brass conduit bush
(d) junior hacksaw
(e) distance saddle

Solution

See Figure 20.

39 Draw the following graphical location symbols:
(a) cooker control unit with 13 A socket outlet
(b) twin tube fluorescent luminaire
(c) twin 13 A switched socket outlet
(d) lighting distribution board
(e) main intake point

Solution

See Figure 21.

40 Draw separate line diagrams for each of the following final circuits using BS 3939 graphical location symbols:
(a) one lighting point controlled by two-way intermediate switching
(b) one immersion heater controlled by a double-pole switch
(c) one 13 A socket outlet ring circuit comprising six sockets on the ring and three spurs at different points on the ring supplying socket outlets
(d) one cooker control point controlling a cooking appliance
(e) two fluorescent luminaires (both twin tubes) controlled by two-way switching

Solution

See Figure 22.

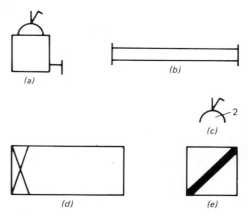

Figure 21

Theory and Regulations

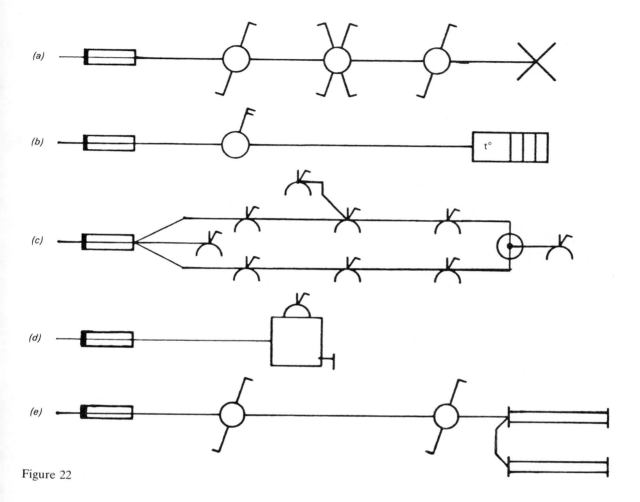

Figure 22

41 From relevant manufacturers' catalogues, select products for the following items, quoting the catalogue number where applicable:
(a) 13 A switched socket outlet
(b) 5 A single-pole, one-way switch
(c) 60 A, 8-way metalclad consumer unit
(d) MCBs – two 5 A, one 15 A, one 20 A, two 30 A and two blank shields
(e) 85 W (1800 mm) single fluorescent luminaire (with diffuser)

Solution

(a) Crabtree 2214 QG.
(b) MK 3591 ALM.
(c) Crabtree 708/2.
(d) Two 50/05; one 50/15; one 50/20; two 50/30 and two 743 blank plates.
(e) Thorn PPQ 675 plus PPC6.

42 Explain the difference between insulation resistance and conductor continuity.

Solution

The difference between the two terms is that *insulation resistance* is concerned with the opposition of leakage current by the insulation medium of circuit cables and components of the wiring system, whereas *conductor continuity* is concerned with the *soundness* of the conducting path taken by current as it flows through the circuit conductors.

Insulation resistance should have a very high ohmic value. The IEE Wiring Regulations require 1 MΩ to be obtained as the minimum insulation resistance value for a test on a completed installation. Insulation resistance of cables is inversely proportional to the length, whereas conductor resistance is proportional to length.

43 A large electrical installation was subdivided into five insulation resistance tests, namely: 0.8 megohm, 100 megohm, 20 megohm, 70 megohm and 2 megohm. What is the equivalent test result of the whole installation? What is an 'outlet' as defined in Regulation 613–5?

Solution

The overall value is found by adding the five test results, remembering that one is dealing with parallel circuits. Hence:

$$\frac{1}{R} = \frac{1}{0.8} + \frac{1}{100} + \frac{1}{20} + \frac{1}{70} + \frac{1}{2}$$

$$= 1.25 + 0.01 + 0.05 + 0.014\,2 + 0.5$$

$$= 1.824$$

$$= \frac{1.824}{1}$$

Therefore:

$$\frac{R}{1} = \frac{1}{1.824} = \mathbf{0.548\ M\Omega}$$

The term 'outlet' refers to points of connection where cables are broken, such as lighting points, socket outlets, switch boxes, etc.

44 Explain how you would identify the cores of a MIMS multicore cable and test it for insulation resistance.

Solution

In practice, the procedure is to use the metal sheath as a return conductor and connect one end of a tester (bell set or ohmmeter) to the sheath and the other end of the tester to individual cores. At the other end of the MIMS cable, cores are separately touched to the sheath and identified accordingly. Individual

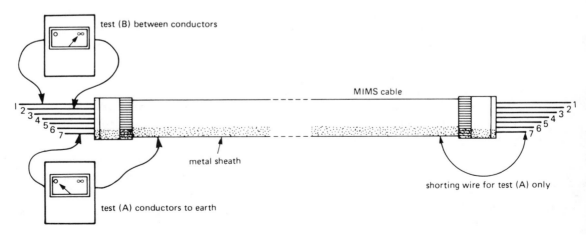

Figure 23

Theory and Regulations

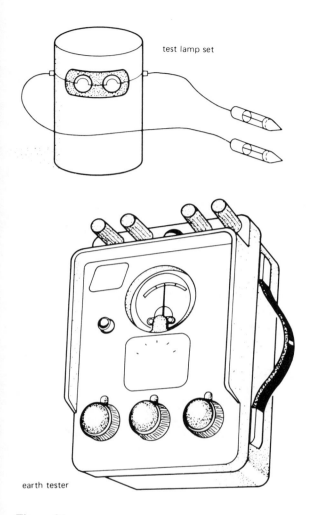

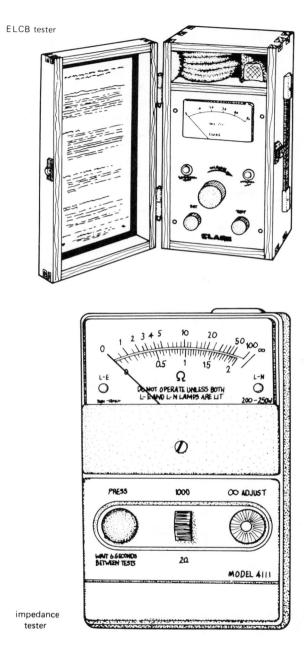

Figure 24

cores are then marked with tape or proper labels to signify their circuit function.

If the tester chosen was an ohmmeter, then the insulation resistance to earth of each core could be ascertained. Care should be taken when using the tester in view of it generating 500 V. In making the test, one should test between cores in case any are touching each other inside the termination glands. Figure 23 shows both methods. It will be noticed that the continuity test is indicating towards zero on the tester and the insulation test indicating towards infinity, meaning that there is an exceedingly high value of resistance between the cores.

45 List the instruments for making the following tests:
  (a) verification of polarity on a live installation
  (b) earth electrode test
  (c) earth leakage circuit breaker test
  (d) earth fault loop impedance test

Solution

(a) Test lamp set.
(b) Null-balance earth tester.
(c) ELCB tester.
(d) Impedance tester.

These instruments are illustrated in Figure 24.

46 What is meant by the *earth-fault-loop path*? Draw a diagram of the path and indicate with arrows the route taken by the fault current.

Solution

The *earth-fault-loop path* is the path taken by escaping current to earth, i.e. earth-leakage current. In Appendix 15 of the IEE Wiring Regulations, the earth fault current loop, starting at the point of fault from the *phase* conductor to earth, passes current into the circuit protective conductor, main earthing terminal and earthing conductor then through the metallic return path of the system (for TN systems) or general mass of earth (for TT and IT systems). The escaping current then passes through the earthed neutral point of the supply transformer and into the faulty phase back to the point of fault. The opposition to this current is known as the *earth-fault-loop impedance* (Z_S).

See Figure 25.

47 What tests would you make on an electrical appliance to find out if it was safe to use?

Solution

The first check that one should do is a visual inspection of the lead and entry of it into the appliance and plug top (or supply source if the appliance is fixed). The inspection of the lead will reveal any wear on the outer sheath, while at entry points one can look for loose cable grips and attachments. Inspection within the plug top and appliance will reveal correct polarity of the flexible lead conductor cores. Look for unnecessary stress on conductors, identification tags (if any), loose termination screws and secure earthing if the appliance is not double insulated. Check size of lead against the rating of the appliance to see if it will adequately carry the load current. Check fuse size and see if any cover screws are missing; also check safety guards (if any).

The appliance needs to be tested for insulation resistance between conductors (phase

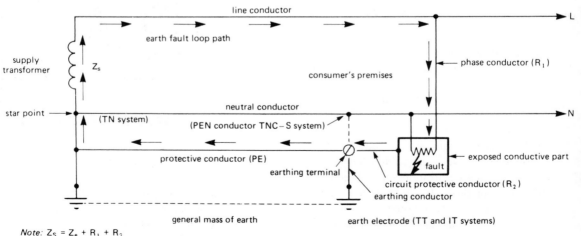

Note: $Z_S = Z_e + R_1 + R_2$

Figure 25

and neutral) and these conductors and earth. This test uses an instrument called an *ohmmeter* and the readings obtained must be in excess of 0.5 megohms. Before this test is applied, however, a test of protective conductor continuity is made, and here the d.c. ohmmeter may be used – see item 3, Appendix 15, IEE Regulations. For further reference, see the description on page 113 and Figure 106 in *Electrical Installation Technology 1*.

48 When should the following installation be inspected?
(a) farms
(b) petrol filling stations
(c) construction sites
(d) dwellings
(e) caravan sites

Solution

Reference should be made to 'Inspection Certificate', page 222, Appendix 16 of the IEE Wiring Regulations.
(a) It is recommended that farms are inspected every three years.
(b) It is often recommended that petrol stations are inspected every year.
(c) For construction sites where temporary supplies are made available, the recommended period between inspections is three months.
(d) It is recommended that dwellings are inspected every five years or less.
(e) The recommended frequency for inspection of caravan sites is every year, but certainly not less than every three years.

49 Draw a circuit diagram of six 13 A single socket outlets connected in the form of a ring final circuit fed from a consumer unit. Illustrate on the diagram a method of testing ring circuit continuity. Explain how you would test socket outlets connected by means of spurs.

Solution

See Figure 26. Socket outlets on spurs are radial circuits. There are no specified test requirements. However, if a test does have to be made, the socket outlet connecting the spur can have its live conductors, in turn, shorted to earth. While at the origin of the spur, a test with an ohmmeter to find the earthed conductor(s) is made.

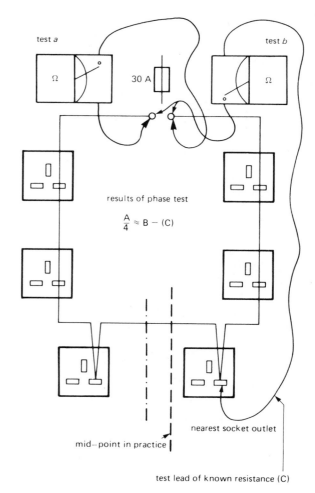

results of phase test

$$\frac{A}{4} \approx B - (C)$$

Figure 26

50 (a) Draw a circuit diagram of a 6-way consumer unit, showing meter tails and outgoing final circuit wiring, and illustrate on the diagram the method(s) of carrying out an insulation resistance test. State the required ohmic values of the test(s).

(b) Show diagrammatically how a polarity test is made on an Edison-type screw lampholder.

Solution

(a) See Figure 27. Reference should be made to the IEE Wiring Regulations, Regs. 613–5, 613–6 and 613–7. Required ohmic value is 1 megohm for completed installation.

(b) See Figure 28.

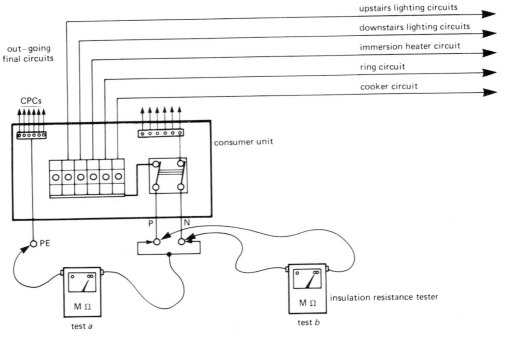

Figure 27

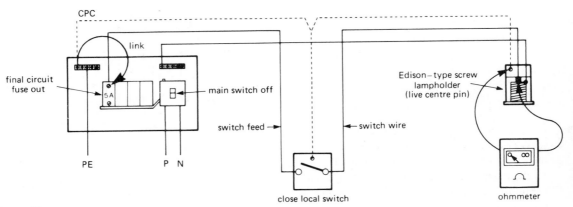

Figure 28

Science and Calculations

51 Write down the unit for:
(a) energy
(b) frequency
(c) resistivity
(d) magnetic flux
(e) luminous flux

Solution

(a) joule (J)
(b) hertz (Hz)
(c) ohm-metre (Ω-m)
(d) weber (Wb)
(e) lumen (lm)

51 What do the following instruments measure?
(a) ammeter
(b) wattmeter
(c) ohmmeter
(d) pyrometer
(e) galvanometer

Solution

(a) current in amperes
(b) power in watts
(c) resistance in ohms
(d) temperature in therms or degrees Celsius
(e) current but sometimes not calibrated in amperes

53 Using metric prefixes simplify the following:
(a) 0.000 5 ampere
(b) 8 954 000 watts
(c) 100 000 ohms
(d) 0.000 000 778 coulomb
(e) 11 000 volts
(f) 15.5 milliamperes
(g) 500 milliwatts
(h) 0.000 600 9 farad
(i) 0.000 333 millihenry
(j) 0.009 8 metre

Solution

(a) Since 1 mA = 1/1 000 A,

then $0.000\ 5\ A = \dfrac{0.000\ 5}{1/1\ 000} = $ **0.5 mA.**

Alternatively, move the decimal point to the right three places, thus: 0.000⁀5A. If the decimal point was moved six places, the answer would be **500 μA**.

(b) Since 1 MW = 1 000 000 W, then
8 954 000 W = **8.954 MW**
(c) 100 000 Ω = **100 kΩ**
(d) 0.000 000 778 C = **778 nC**
(e) 11 000 V = **11 kV**
(f) **15.5 mA**
(g) **500 mW**
(h) 0.000 600 9 F = **600.9 μF**
(i) 0.000 333 mH = **333 nH**
(j) 0.009 8 m = **9.8 mm**

54 Express the following in terms of their numerical values, writing after each the appropriate unit symbol – as in the first example:
  (a) twenty-four thousand milliamperes – 24 000 mA
  (b) forty-three megohms
  (c) eight hundred thousand microwatts
  (d) two million voltamperes
  (e) fifty hertz
  (f) four hundred and fifteen volts
  (g) four and three-quarter litres
  (h) one half microhms
  (i) eleven kilovolts
  (j) forty-four milliseconds

Solution

  (a) Note that 24 000 mA = **24 A**
  (b) 43 000 000 Ω = **43 MΩ**
  (c) 800 000 μW = **800 mW**
  (d) 2 000 000 VA = **2 MVA**
  (e) **50 Hz**
  (f) **415 V**
  (g) **4.75 l**
  (h) **0.5 μΩ**
  (i) **11 kV**
  (j) **44 ms**

55 Simplify the following values using prefix symbols:
  (a) 1 850 W
  (b) 0.123 A
  (c) 0.005 005 J
  (d) 40 992 143 s
  (e) 0.035 8 V

Solution

  (a) 1.85 kW
  (b) 123 mA
  (c) 5 005 μJ
  (d) 40.99 Ms
  (e) 35.8 mV

56 Rewrite the following using symbols to replace the words in italics (i.e. slanting words).
  (a) 250 MW is *greater than* 250 kW
  (b) 45.59 J is *approximately equal to* 0·0456 kWs
  (c) *efficiency* = 79%
  (d) the constant of a circle is *pi*
  (e) 1 microfarad *equals* 0.000 001 farad

Solution

  (a) 250 MW > 250 kW
  (b) 45.59 J ≈ 0.0456 kWs
  (c) η = 79%
  (d) the constant of a circle is π
  (e) 1 microfarad = 0.000 001 farad

57 Complete the following table by applying Ohm's Law:

V	240 V		10 V	50 mV		50 kV
I		25 mA	2 kA		0.02 A	0.01 A
R	19.2 Ω	5 Ω		1 μΩ	0.002 Ω	

Solution

I = 12.5 A
V = 125 mV
R = 5 mΩ
I = 50 kA
V = 40 μV
R = 5 MΩ

58 Complete the following table using the impedance right-angled triangle for a.c. circuits (see Figure 35):

Z	50 Ω		100 Ω	20 Ω		$\sqrt{3}$ Ω
X		10 Ω	60 Ω		64 Ω	$\sqrt{2}$ Ω
R	30 Ω	10 Ω		12 Ω	32 Ω	

Solution

$X = 40\ \Omega$
$Z = 14.14\ \Omega$
$R = 80\ \Omega$
$X = 16\ \Omega$
$Z = 71.55\ \Omega$
$R = 1\ \Omega$

59 The energy used by a resistive load in 14 hours is 847 MWh. What is the current taken by the load if the supply voltage is 11 kV?

Solution

In this question, first divide the 14 hours into the 847 MWh (megawatt-hours). Thus:

$$P = \frac{\text{MWh}}{\text{h}} = \frac{847}{14} = \mathbf{60.5\ MW}$$

From this:

$$I = \frac{P}{V} = \frac{60.5 \times 10^6}{11 \times 10^3} = \mathbf{5\ 500\ A}$$

60 A cable has a total resistance of 0.6 Ω. When it carries a current of 60 A, determine:
(a) its volt drop
(b) its power loss
(c) the energy consumed over 24 hours

Solution

(a) As pointed out earlier, volt drop is found by:
$V = IR = 60 \times 0.6 = \mathbf{36\ V}$
(b) Power loss is found by:
$P = I^2R = 60 \times 60 \times 0.6 = 2\ 160\ \text{W or}$ **2.16 kW**
(c) Energy used is found by:
$W = Pt = 2.16 \times 24 = \mathbf{51.84\ kWh}$

61 Which of the following conductors has the least electrical resistance?
(a) a short copper conductor of 2.5 mm²
(b) a short copper conductor of 1.5 mm²
(c) a long copper conductor of 1.5 mm²
(d) a long copper conductor of 2.5 mm²

Solution

This is a relatively easy question if students can remember resistance and its dependence on dimensions (see 'Resistance factors', page 29, *Electrical Installation Technology 2*). The conductor with the least electrical resistance will be the shortest and thickest one, that is **(a)**.

62 Calculate the total current of six 40 W, 240 V lamps connected in parallel.

Solution

This is another simple question based on the power expression used before, i.e. $P = VI$. From this:

$$I = \frac{P}{V} = \frac{6 \times 40}{240} = \mathbf{1\ A}$$

Note that each lamp takes 0.166 A.

63 Calculate the total wattage of six 100 W, 240 V lamps connected (a) in series and (b) in parallel.

Solution

(a) In this arrangement, each lamp does not receive its full working voltage, but only one-sixth of the supply voltage, i.e. 40 V. Since each lamp has a working resistance of:

$$R = \frac{V^2}{P} = \frac{240 \times 240}{100} = 576\ \Omega$$

then power of each lamp is:

$$P = \frac{V^2}{R} = \frac{40 \times 40}{576} = 2.777\ W$$

total wattage is: $6 \times 2.777 = \mathbf{16.66\ W}$

(b) In this arrangement, each lamp is connected across the 240 V supply and all lamp wattages can be added together, i.e. **600 W**. Notice that:

$$\frac{600}{16.66} = 36 \text{ also } \frac{240}{40} = 6.$$

The voltage squared is an important factor since $6^2 = 36$. Arrangement (b) gives 36 times more power.

64 The resistance of a 60 W lamp is 960 Ω. What is the current taken by the lamp?

Solution

Since $P = I^2R$ then:

$$I = \sqrt{\frac{P}{R}} = \sqrt{\frac{60}{960}} = \mathbf{0.25\ A}$$

Further calculation will show that the lamp is 240 V.

65 The primary winding of a transformer consists of 400 turns, its secondary voltage is 240 volts and its primary current is 8 amperes. Determine the primary voltage, secondary turns and secondary current if its transformation ratio is 1:1.73.

Solution

First, students must remember that the transformation ratio of a transformer is given by:

$$\frac{V_p}{V_s} = \frac{N_p}{N_s} = \frac{I_s}{I_p}$$

where V_p is primary volts
V_s is secondary volts
N_p is primary turns
N_s is secondary turns
I_p is primary current
I_s is secondary current

Since each ratio = 1:1.73 (step-up)

then

$$\frac{V_p}{240} = \frac{1}{1.73}$$

therefore

$$V_p = \frac{240}{1.73}$$

$$= \mathbf{138.7\ V}$$

Similarly:

$$\frac{400}{N_s} = \frac{1}{1.73}$$

and

$$N_s = 400 \times 1.73$$
$$= \mathbf{692\ turns}$$

Also:

$$\frac{I_s}{8} = \frac{1}{1.73}$$

and

$$I_s = \frac{8}{1.73}$$

$$= \mathbf{4.62\ A}$$

66 A transformer has a turns ratio of 250:20. What is the secondary current if the primary current is 10 A?

Solution

The first thing to note is that the transformer is a step-down one with 250 turns on its primary and 20 turns on its secondary. The transformation ratio is 12.5:1 and from the previous question:

$$\frac{I_s}{I_p} = 12.5$$

Thus:

$$I_s = I_p \times 12.5$$
$$= 10 \times 12.5$$
$$= \mathbf{125\ A}$$

Science and Calculations

67 An auto-transformer has 480 turns on its primary winding. If it is supplied with 240 volts and its secondary voltage is 24 volts, how many turns has its secondary winding?

Solution

As indicated, an auto-transformer is one with only one winding as shown in Figure 29. It will be noticed that the secondary voltage is from a tapping point on the winding and since the ratio:

$$\frac{V_p}{V_s} = \frac{240}{24} = 10$$

then

$$\frac{N_p}{N_s} = 10$$

therefore:

$$N_s = \frac{N_p}{10} = \frac{480}{10} = \textbf{48 turns}$$

68 Describe the principles of transformer action, and write down the expression for the transformation ratio.

Solution

The answer to this question can be found in *Electrical Installation Technology 2*, pages 90–91. Briefly, a double-wound transformer operates on the principle of mutual induction, i.e. when an a.c. supply is fed into the primary winding of the transformer, the current that flows produces an alternating magnetic flux in the iron core. This flux induces not only a back e.m.f. in the primary winding but also an e.m.f. in the secondary winding. The voltage induced in the secondary winding depends upon the number of turns of this winding. Since both windings are linked by the same magnetic flux, their induced e.m.f.s will be proportional to the number of turns in each coil.

69 Determine the power factor of an inductive circuit supplied at 240 V and taking a current of 30 A. Assume the load has an output of 4.5 kW and efficiency of 86%.

Solution

Reference should be made to page 61 of *Electrical Installation Technology 2* where power factor is explained. It will be seen that:

$$\text{p.f.} = \frac{P}{VI}$$

where P in the input power in watts

Since efficiency = $\frac{\text{output}}{\text{input}}$

the input power = $\frac{\text{output}}{\text{efficiency}}$

$$= \frac{4\,500}{86} \times 100$$

$$= 5233 \text{ W (approx.)}$$

therefore

$$\text{p.f.} = \frac{5233}{240 \times 30}$$

$$= \textbf{0.727 lagging}$$

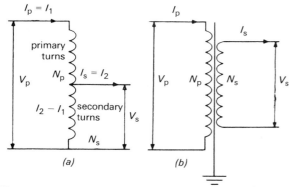

Figure 29

70 (a) Explain the following terms: frequency; periodic time; maximum value; r.m.s. value.
(b) With the aid of a diagram, explain how a.c. is produced from a simple single loop generator.

Solution

(a) *frequency* This describes the number of cycles completed in a time of one second (usually). Its derived SI unit is called the *hertz*.
periodic time This is the time taken to complete one cycle and the relationship between itself and frequency is given by the expression:

$$T = \frac{1}{f}$$

maximum value This is the highest point (instantaneous point) reached by an a.c. waveform often called the *peak* value.
r.m.s. value This is called the *root mean square* value or *effective* value of an a.c. waveform, be it voltage or current. Its value can be derived from the expression:

$$V_{r.m.s.} = 0.707 \times V_{max}$$

or $I_{r.m.s.} = 0.707 \times I_{max}$

(b) The answer to this part can be found on page 39 of *Electrical Installation Technology 2*. Briefly, the loop shown in Figure 30 has to be rotated for the conductor to produce an induced e.m.f. This is brought about by *flux cutting* as the conductor rotates through the magnetic field. Current in the external circuit will only flow if the circuit is complete through some form of load. Apply Fleming's right-hand rule to ascertain the current direction.

71 With reference to Figure 31 determine the following:
(a) equivalent circuit resistance
(b) current flowing in the circuit
(c) quantity of electricity over a period of 6 hours
(d) potential difference across each resistor
(e) power consumed by each resistor and total power
(f) energy used in the circuit over a period of 6 hours.

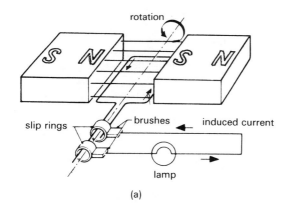

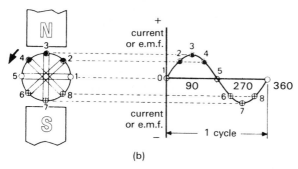

Figure 30 *(a) Single loop generator (b) Rotating conductor*

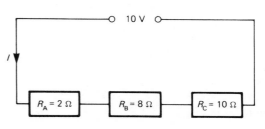

Figure 31

Solution

(a) $R_e = R_A + R_B + R_C$
$= 2 + 8 + 10$
$= \mathbf{20\ \Omega}$

(b) $I = \dfrac{V}{R_e}$
$= \dfrac{10}{20}$
$= \mathbf{0.5\ A}$

(c) $Q = I \times t$
$= 0.5 \times 6 \times 3\ 600$
$= \mathbf{10\ 800\ C}$

(d) $V_A = I \times R_A$
$= 0.5 \times 2$
$= \mathbf{1\ V}$
$V_B = I \times R_B$
$= 0.5 \times 8$
$= \mathbf{4\ V}$
$V_C = I \times R_C$
$= 0.5 \times 10$
$= \mathbf{5\ V}$

(e) $P_A = V_A \times I$
$= 1 \times 0.5$
$= \mathbf{0.5\ W}$
$P_B = V_B \times I$
$= 4 \times 0.5$
$= \mathbf{2.0\ W}$
$P_C = V_C \times I$
$= 5 \times 0.5$
$= \mathbf{2.5\ W}$
$P = P_A + P_B + P_C$
$= 0.5 + 2.0 + 2.5$
$= \mathbf{5\ W}$

(f) $W = P \times t$
$= 5 \times 6 \times 3\ 600$
$= \mathbf{108\ 000\ J}$

72 From the answers given in Question 71 express:
(a) equivalent resistance in terms of kilo-ohms
(b) current in terms of milliamperes
(c) quantity of electricity in terms of megacoulombs
(d) energy in terms of megajoules and kilowatt-hours

Solution

(a) Since $1\ k\Omega = 1\ 000\ \Omega$
then $20\ \Omega = \dfrac{20}{1\ 000} = \mathbf{0.02\ k\Omega}$

(b) Since $1\ mA = \dfrac{1}{1\ 000}\ A$
then $0.5\ A = \dfrac{0.5}{1/1\ 000} = \mathbf{500\ mA}$

(c) Since $1\ MC = 1\ 000\ 000\ C$
then $10\ 800\ C = \dfrac{10\ 800}{1\ 000\ 000} = \mathbf{0.0108\ MC}$

(d) Since $1\ MJ = 1\ 000\ 000\ J$
then $108\ 000\ J = \dfrac{108\ 000}{1\ 000\ 000} = \mathbf{0.108\ MJ}$

Also:
since $1\ kWh = 3.6\ MJ$
then $0.108\ MJ = \dfrac{0.108}{3.6} = \mathbf{0.03\ kWh}$

73 With reference to Figure 32 determine the following:
(a) equivalent circuit resistance
(b) value of each resistor

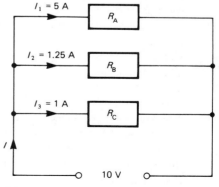

Figure 32

(c) power consumed by each resistor and total power
(d) energy used in the circuit over a period of 6 hours

Solution

(a) $R_e = \dfrac{V}{I} = \dfrac{10}{7.25} = \mathbf{1.379\ \Omega}$

(b) $R_A = \dfrac{V}{I} = \dfrac{10}{5} = \mathbf{2\ \Omega}$

$R_B = \dfrac{V}{I} = \dfrac{10}{1.25} = \mathbf{8\ \Omega}$

$R_C = \dfrac{V}{I} = \dfrac{10}{1} = \mathbf{10\ \Omega}$

(c) $P_A = I^2R = 5 \times 5 \times 2 = \mathbf{50\ W}$
$P_B = I^2R = 1.25 \times 1.25 \times 8 = \mathbf{12.5\ W}$
$P_C = I^2R = 1 \times 1 \times 10 = \mathbf{10\ W}$
Total power $(P) = P_A + P_B + P_C$
$= 50 + 12.5 + 10$
$= \mathbf{72.5\ W.}$
(Note that $P = V \times I = 10 \times 7.25$
$= 72.5\ W$)

(d) $W = P \times t = 72.5 \times 6 \times 3\ 600$
$= 1\ 566\ 000\ J = \mathbf{1.566\ MJ}$

74 With reference to Questions 71 and 73, state several comparisons between both circuits.

Solution

(a) In the series circuit the current is common to all the resistors, whereas in the parallel circuit it is the supply voltage which is common across all the resistors.
(b) In the series circuit the sum of the internal p.d.s. add up to the supply voltage, whereas in the parallel circuit it is the sum of the branch currents which adds up to the total current flowing in the circuit.
(c) The equivalent resistance of the parallel circuit is smaller than the smallest resistor connected, whereas in the series circuit the total resistance is the sum of the individual resistors.

(d) Power consumption is greater in the parallel circuit than in the series circuit.

75 Three resistors of 8 Ω, 12 Ω, 24 Ω respectively, are connected across a 220 V supply. Determine the equivalent resistance of the group: (a) in series and (b) in parallel. In each case, find the power consumed.

Solution

(a) $R = R_1 + R_2 + R_3$
$= 8 + 12 + 24 = \mathbf{44\ \Omega}$

$P = \dfrac{V^2}{R} = \dfrac{220 \times 220}{44} = \mathbf{1.1\ kW}$

(b) $\dfrac{1}{R} = \dfrac{1}{R_1} + \dfrac{1}{R_2} + \dfrac{1}{R_3}$

$\dfrac{1}{8} + \dfrac{1}{12} + \dfrac{1}{24} = \dfrac{6}{24}$

$\therefore R = \mathbf{4\ \Omega}$

$P = \dfrac{V^2}{R} = \dfrac{220 \times 220}{4} = \mathbf{12.1\ kW}$

76 Calculate the power dissipated by a 110 V filament lamp having a working resistance of 121 Ω.

Solution

In practice, lamp wattage is stated on the outer glass envelope of most lamps. While the calculation of power is quite simple, using the expression in Question 75; it should be noted that the lamp's resistance is only 40 ohms at normal room temperature, i.e. about 25 °C.

$P = \dfrac{V^2}{R} = \dfrac{110 \times 110}{121}$
$= \mathbf{100\ W}$

77 Determine the supply voltage to a 150 W projector lamp if it takes a current of 0.625 A. Explain what would happen if the supply voltage was increased by 2.5%.

Solution

In this question:

$$V = \frac{P}{I} = \frac{150}{0.625} = \mathbf{240\ V.}$$

The maximum voltage variation allowed on a consumer's premises is ± 6% of the declared voltage. However, if the voltage increased by 2.5%, the power of the lamp would increase.

Since

$$P = \frac{V^2}{R}$$

where

$$R = \frac{V}{I} = \frac{240}{0.625} = 384\ \Omega$$

and

$$V = (2.5\%\ \text{of}\ 240\ V) + 240\ V = 246\ V$$

Hence

$$P = \frac{246 \times 246}{384}$$

$$= \mathbf{157.6\ W}$$

It should be pointed out that lamp manufacturers do not like to see any marked increase in supply voltage since this reduces the expected lamp life.

78 The insulation resistance of a certain cable is 500 MΩ. What leakage current is likely to flow when the cable's insulation is subject to a stress voltage of 500 V?

Solution

The cable has a very good insulation resistance value and consequently the leakage current is very small, i.e.

$$I = \frac{V}{R} = \frac{500}{500} \times 10^{-6}$$

$$= 0.000\ 001\ A\ \text{or}\ \mathbf{1\ \mu A}$$

It should be noted that the longer the cable run, the less its insulation resistance value will be: its value is inversely proportional to its length.

79 A battery consisting of nine primary cells is connected to an external resistance of 10 Ω. If each cell has an e.m.f. of 1.5 V and internal resistance 0.45 Ω, determine the circuit current and volt drop across the 10 Ω resistor when the cells are arranged in (a) series, (b) parallel and (c) three sets in parallel, each consisting of three cells in series.

Solution

Figure 33 illustrates the circuit connections.

(a) $$I = \frac{nE}{R + nr}$$

where n is the number of cells
R is the external resistor
r is the internal resistance of each cell

Thus $$I = \frac{9 \times 1.5}{10 + (9 \times 0.45)}$$

$$= \mathbf{0.961\ A}$$

Also $V = I \times R$
$= 0.961 \times 10$
$= \mathbf{9.61\ V}$

(b) $$I = \frac{E}{R + r/9}$$

$$= \frac{1.5}{10 + 0.05}$$

$$= \mathbf{0.149\ A}$$

$V = I \times R$
$= 0.149 \times 10$
$= \mathbf{1.49\ V}$

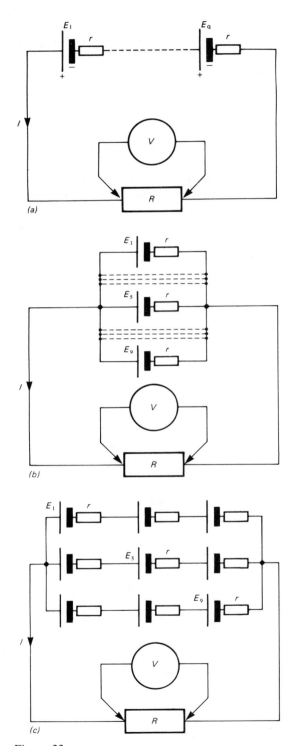

(c) $I = \dfrac{n/3 \times E}{R + r}$

where $r = \dfrac{3 \times 0.45}{3}$

$I = \dfrac{3 \times 1.5}{10 + 0.45}$

$= \mathbf{0.43\ A}$

$V = I \times R$
$= 0.43 \times 10$
$= \mathbf{4.3\ V}$

Note: The series cell connections provide the highest battery voltage, but when current flows a high internal volt drop occurs (approximately 4 V). The parallel arrangement produces the lowest internal volt drop, only 0.01 V, but suffers by having a low battery voltage. In the series/parallel connection a compromise is found; battery voltage is reasonable and the internal volt drop of 0.2 V is due to the internal resistance being equivalent to one cell.

80 Briefly state the main differences between Planté, tubular and flat plate lead-acid cells.

Solution

The main differences between the cells are in the positive plates, the containers which are used and also their expected life.

The Planté positive plate is made of pure lead. A number of thin vertical lamelles provide a large active surface area which gives a high capacity and continuous regeneration of active material from the basic metal. This type of cell gives 100% capacity throughout its very long life. The containers used for Planté cells are transparent styrene acrylonitrite in order to allow visual inspection to be made of electrolyte levels.

In the tubular positive plate type, the tubes are made from terylene or other similar material fitted over cast lead alloy spines attached to a common metallic busbar at the top of the plate. The space between the lead spine and the tubing is filled with lead oxide

Figure 33

which is retained within the tube by a plastic plug. In this way the active material is maintained within the conducting spines during the cell's charge and discharge. In practice, these cells give high power for a minimum volume and are very robust. Containers may also be of transparent plastic for maintenance reasons but the cell's life expectancy is not as good as the Planté type.

In the flat plate type cells, lead oxide paste is pressed into a metallic grid which serves the purpose of retaining the active material as well as a conducting medium for the passage of current. The positive plate of these pasted cells is similar to the negative plate of the Planté and tubular cells but their containers may again be transparent to allow visual inspection to be made. For harder use, containers are available made from hard rubber.

81 Determine the efficiency of an electric kettle rated at 3kW/240 V when it contains 1 litre of water. The change of temperature from cold to boiling point is 80 °C and the time taken to boil is 2 minutes 10 seconds. Assume the specific heat capacity of water to be 4.2 kJ/kg °C and that 1 litre = 1 kg.

Solution

Heat energy required (output):
$$W_0 = mc(\theta_2 - \theta_1)$$
$$= 1 \times 4200 \times 80$$
$$= 0.336 \text{ MJ}$$
since 1 kWh = 3.6 MJ then
$$W_0 = \frac{0.336}{3.6} \text{ kWh}$$
$$= 0.093 \text{ kWh}$$

Heat energy required (input):
$$W_i = Pt$$
$$= 3 \times \frac{130}{3\,600}$$
$$= 0.108 \text{ kWh}$$

$$\% \text{ efficiency} = \frac{\text{output}}{\text{input}} \times 100$$
$$= \frac{0.093 \times 100}{0.108}$$
$$= 86\%$$

82 Figure 34 shows the connections of four capacitors, determine:
(a) the equivalent capacitance of the circuit
(b) the total charge for the p.d. given
(c) the total energy stored

Solution

(a) In the parallel branch the equivalent capacitance is $C = C_1 + C_2$.
Thus $C = 5 + 5 = 10 \text{ }\mu F$
The equivalent capacitance of the whole circuit is:
$$\frac{1}{C_e} = \frac{1}{C} + \frac{1}{C_3} + \frac{1}{C_4}$$
$$= \frac{1}{10} + \frac{1}{5} + \frac{1}{5}$$
$$= \frac{1 + 2 + 2}{10}$$
$$= \frac{5}{10} \mu F$$
$$C_e = \frac{10}{5}$$
$$= 2 \text{ }\mu F$$

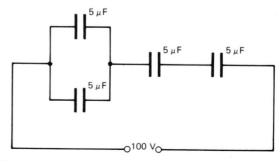

Figure 34

(b) The total charge is given by the expression $Q = CV$
Thus $Q = 2 \times 10^{-6} \times 100$
= **200 μC**

(c) The total energy is found by the expression $W = \tfrac{1}{2}CV^2$
Thus $W = \tfrac{1}{2} \times 2 \times 10^{-6} \times 100 \times 100$
= **0.01 J**

83 Figure 35 represents an impedance triangle.
(a) Write down an expression for finding Z and determine its ohmic value.
(b) State meanings for the following terms: impedance; reactance and resistance.

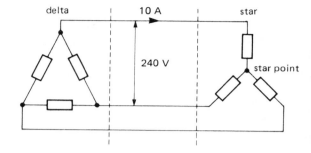

Figure 36

Solution

(a) $Z = \sqrt{R^2 + X^2}$
$Z = \sqrt{20^2 + 40^2}$
 $= \sqrt{400 + 1\,600}$
 $= \sqrt{2\,000}$
 = **44.72 Ω**

(b) *Impedance* This is the ratio of voltage to current (V/I) in r.m.s. terms for a.c. quantities. Since it is represented by the hypotenuse in Figure 35 it will be the largest value in the circuit or total opposition to current flow.
Reactance In a.c. circuits it is the opposition to current flow by components possessing inductance or capacitance such as stator windings, transformers, lamp ballasts (which possess inductance) or capacitors (which possess capacitance).
Resistance The property of a material to resist the flow of current through a circuit such as a resistor.

84 With reference to the balanced system in Figure 36 determine:
(a) the phase current in the delta connection
(b) the phase current in the star connection
(c) the phase voltage in the delta connection
(d) the phase voltage in the star connection

Solution

(a) $I_P = \dfrac{I_L}{\sqrt{3}} = \dfrac{10}{1.732} =$ **5.77 A**

(b) $I_P = I_L =$ **10 A**

(c) $V_P = V_L =$ **240 V**

(d) $V_P = \dfrac{V_L}{\sqrt{3}} = \dfrac{240}{1.732} =$ **138.6 V**

Note: I_P is phase current
I_L is line current
V_P is phase voltage
V_L is line voltage
$\sqrt{3} = 1.732$

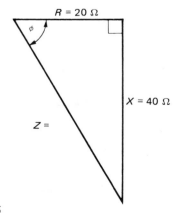

Figure 35

85 (a) With reference to Figure 35, the angle (ϕ) shown can represent power factor. If the reactance is inductive, determine the power factor, stating whether it is *unity*, *lagging* or *leading*.
(b) Write down a more exact expression for finding power factor and provide a brief explanation as to its importance in a.c. circuits.

Solution

(a) Having already worked out Z to be 44.72 Ω, power factor ($\cos \phi$) can be determined using trigonometry, i.e.:

$$\cos \phi = \frac{\text{adjacent}}{\text{hypotenuse}} = \frac{R}{Z} = \frac{20}{44.72}$$

$$= 0.447$$

Hence power factor is 0.447 lagging (since reactance is inductive).

(b) Power factor is the ratio of

$$\frac{\text{active power } (P)}{\text{apparent power}(VI)}$$

An explanation of power factor is given in *Electrical Installation Technology 2*, page 53.

The active power and voltage in a system are often fixed, consequently any power factor less than unity (1) will cause more current to flow in the system than otherwise necessary. For this reason, supply authorities penalize consumers who have low power factors, i.e. those below 0.8 lagging, because it means that larger cables and switchgear need to be installed to meet the consumer's load conditions.

86 (a) Show by diagram how an induced e.m.f. is created using a permanent magnet, coil and millivoltmeter.
(b) State a number of principles associated with (a) above.
(c) What is the value of induced e.m.f. in a conductor 0.4 m long if it moves at a velocity of 20 m/s at right angles to a magnetic field of strength 2.5 T?

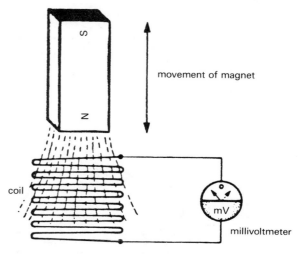

Figure 37

Solution

(a) See Figure 37.
(b) (i) An e.m.f. is induced if either the magnet or the coil is moved relative to each other.
(ii) The magnitude of the induced e.m.f. depends on the rate at which either the magnet or coil is moved towards each other.
(iii) The polarity of the induced e.m.f. depends on the polarity of the permanent magnet and on the direction of movement.
(c) An expression for finding induced e.m.f. is $e = Blv$
where B is the magnetic flux density in tesla (T)
l is the effective length in metres (m)
v is the velocity in metre/second (m/s)
Thus $e = 2.5 \times 0.4 \times 20$
= **20 V**

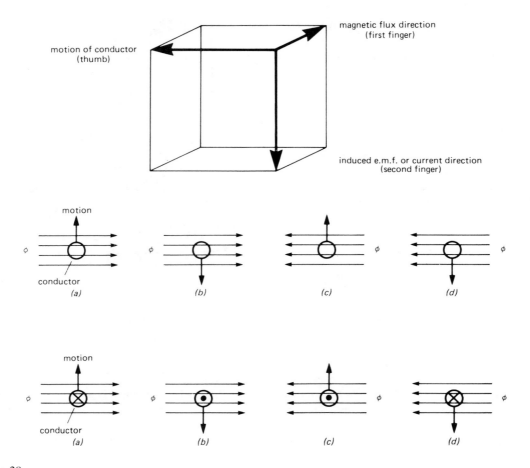

Figure 38

87 Using the model of Fleming's right-hand rule in Figure 38, determine the induced current directions in the four conductors.

Solution

The application of the rule is as suggested, the induced current direction shown by a *cross* denoting current going in and a *dot* denoting current coming out (see page 36, *Electrical Installation Technology 2*).

88 Write down an expression for finding:
(a) *current* when given *charge* and *time*
(b) *voltage* when given *power* and *resistance*
(c) *flux density* when given *force*, *current* and *length*
(d) *volt drop* when given *current* and *resistance*
(e) *resistivity* when given *resistance*, *length*, and *cross-sectional area*

Solution

(a) since $Q = It$ then $I = Q/t$
(b) since $P = VI$ and $I = V/R$ then $P = VV/R$
and $V = \sqrt{PR}$
(c) since $F = Bil$ then $B = F/il$
(d) $V = IR$
(e) since $R = \dfrac{pl}{a}$ then $\varrho = Ra/l$

Note: See pages 14 and 15, *Electrical Installation Technology 2.*

89 (a) With the aid of a diagram illustrate how a conductor carrying current is made to move out of the influence of a permanent magnetic field.
(b) A conductor 50 mm in length lies at right angles to a magnetic field of field strength 25 T. Calculate the force on the conductor when it carries a current of 10 A.
(c) Give *two* practical examples of (a) above.

Solution

(a) See Figure 39.
(b) Force on the conductor is given by:
$F = B \times l \times I$ newtons
where B is the magnetic field strength (T)
I is the current in the conductor (A)
l is the length of conductor (m)
thus $F = 25 \times 0.05 \times 10$
   $= \mathbf{12.5\ N}$

(c) Motor operation
Instrument operations

90 A length of conduit 3 m long is cut into three pieces, the second piece is twice as long as the first while the third piece is 40 cm shorter than the first. If 20 cm is allowed for wastage, find the length of the three pieces.

Solution

The solution to this question can be found using algebra, taking the first piece as x, the second piece as $2x$ and the third piece as $x - 40$. The total length of these pieces is 20 cm short of 3 m, i.e. 280 cm. Thus

$$x + 2x + (x - 40) = 280$$
$$4x - 40 = 280$$
$$4x = 320$$
$$\text{First piece } x = \frac{320}{4}$$
$$= \mathbf{80\ cm}$$
$$\text{Second piece } 2x = \mathbf{160\ cm}$$
$$\text{Third piece } x - 40 = \mathbf{40\ cm}$$

91 Find the volume of a round and hollow copper busbar 15 m in length having internal and external diameters of 5.5 cm and 6.0 cm respectively.

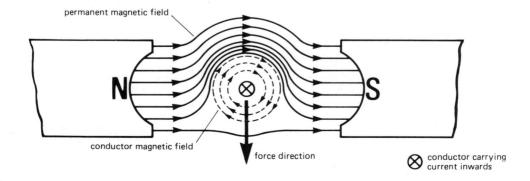

Figure 39

Solution

Volume = area of end (thickness of tube) × length

$$= \left(\pi \frac{d_1^2}{4} - \pi \frac{d_2^2}{4}\right) \times L$$

where $d_1 = 6.0$ cm
$d_2 = 5.5$ cm

$$\text{Volume} = \frac{\pi L}{4}(d_1^2 - d_2^2)$$

$$= \frac{\pi L}{4}(d_1 + d_2)(d_1 - d_2)$$

$$= \frac{3.142 \times 1\,500 \times 11.5 \times 0.5}{4}$$

$$= \mathbf{6\,775\ cm^3}\ (\text{approx.})$$
$$= \mathbf{0.006\,775\ m^3}$$

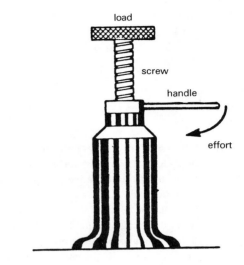

Figure 40

92 Six 125 W, 240 V fluorescent discharge lamps are to be connected into a lighting distribution fuseboard. Determine:
(a) the load current of the circuit (assuming the 1.8 factor)
(b) the size of fuse protecting the circuit
(c) the current rating of the control circuit switches

Solution

(a) $I = \dfrac{P \times 1.8}{V}$

$= \dfrac{125 \times 6 \times 1.8}{240}$

$= \mathbf{5.6\ A}$ (approx.)

(b) Size of fuse (BS 3036) is **15 A**
(c) Current rating of control switches is **15–20 A**. See Regulation 537–19 (switches for discharge lighting circuits)

93 Determine the assessed current demand of three 12 kW/240 V cooking appliances used in a boarding house.

Solution

References should be made to Table 4B, Appendix 4, IEE Wiring Regulations. Here it will be seen that there is no diversity allowed on the first cooking appliance, an 80% allowance on the second and 60% allowance on remaining appliances. Thus:

First appliance

$$I = \frac{P}{V} = \frac{12\,000}{240} = 50\ A$$

Second appliance

$$I = \frac{P}{V} = \frac{12\,000 \times 0.8}{240} = 40\ A$$

Third appliance

$$I = \frac{P}{V} = \frac{12\,000 \times 0.6}{240} = 30\ A$$

The assessed current demand is therefore
50 + 40 + 30 = **120 A**

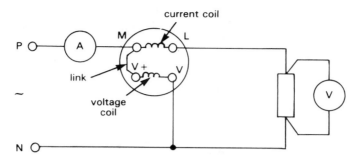

Figure 41

94 (a) Make a sketch of a simple screwjack.
 (b) The screw of a screwjack has 500 threads per metre. If its handle is 0.25 m in length what is its movement ratio?
 (c) If the screwjack had an efficiency of 32%, what load could be lifted for an applied effort of 95 N?

Solution

(a) See Figure 40.
(b) movement ratio (MR)

$$= \frac{2\pi \times \text{length of handle}}{\text{screw pitch}}$$

$$= \frac{2 \times 3.142 \times 0.25}{1/500}$$

$$= \frac{1.571}{0.002}$$

$$= \mathbf{785.5}$$

(c) load = effort × force ratio (FR) and FR = MR × efficiency

Thus load = $95 \times 785.5 \times \frac{32}{100}$

$$= \mathbf{23.88 \text{ kN}}$$

95 Draw a circuit diagram showing how a single-phase wattmeter, voltmeter and ammeter are connected to a resistive load.

Solution

See Figure 41.

96 Determine the efficiency and power factor of a single-phase motor having the following data:
Electrical input – wattmeter reading 20 kW
 – voltmeter reading 240 V
 – ammeter reading 100 A
Mechanical output – 16 kW

Solution

% Efficiency = (output × 100)/input
 = 1600/20
 = **80%**
Power factor = input power/voltamperes
 = 20 000/(240 × 100)
 = **0.833 lagging**

97 From Figure 42, determine the value of the resistor marked X if the p.d. across the 2 Ω resistor is 60 V.

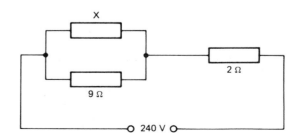

Figure 42

Solution

The current flowing through the circuit is found by:

$$I = \frac{V_2}{R} = \frac{60}{2} = 30 \text{ A}$$

Since $E = V_1 + V_2$

where E is the supply voltage

V_1 and V_2 are the circuit p.d.s

then $V_1 = E - V_2$
$= 240 - 60$
$= 180 \text{ V}$

The current through the 9 Ω resistor is:

$$I = \frac{V_1}{R} = \frac{180}{9} = 20 \text{ A}$$

The current through the unknown resistor is:
$30 - 20 = 10 \text{ A}$

The value of the unknown resistor is:

$$R_X = \frac{V_1}{I} = \frac{180}{10} = 18 \text{ Ω}$$

98 Find the time required for an immersion heater to raise the temperature of 30 litres of water from 20 °C to 70 °C if the heating element is 3 kW, 240 V and the heating system is 82% efficient. Assume the heat specific capacity of water to be 4180 J/kg °C.

Solution

$$\text{heat energy required} = mc\,(\theta_2 - \theta_1)$$
$$= 30 \times 4\,180 \times 50$$
$$= 6.27 \times 10^6 \text{ J}$$

$$\text{electrical energy required} = \frac{6.27 \times 10^6 \times 100}{82}$$
$$= 7.646 \times 10^6 \text{ J}$$

$$\text{time} = \frac{\text{electrical energy}}{\text{power}}$$
$$= \frac{7.646 \times 10^6}{3\,000}$$
$$= 2\,548.8 \text{ s}$$
$$= 43 \text{ mins (approx.)}$$

99 A load of 60 kg is placed 100 mm from the fulcrum of a lever. What force is required on the other side of the lever which is 0.4 m from the fulcrum in order to raise the load?

Solution

This question relates to the *principle of moments* which states that in any system which is in equilibrium (i.e. balance), the total clockwise moment equals the total anticlockwise moment.

Thus

$$F_1 \times d_1 = F_2 \times d_2$$

and

$$F_1 = \frac{F_2 \times d_2}{d_1}$$

where F_1 is the force of the clockwise movement

d_1 is the distance of the clockwise moment

F_2 is the force of the anticlockwise moment

d_2 is the distance of the anticlockwise moment

See Figure 43.

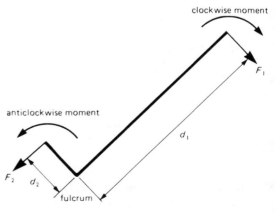

Figure 43

Science and Calculations

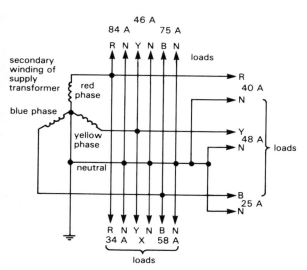

Figure 44

therefore

$$F_1 = \frac{60 \times 100}{400} \times 9.81$$

$$= 147.15 \text{ N}$$

100 (a) Why is it important to balance single-phase a.c. loads on a three-phase, four-wire supply system?

(b) What current in Figure 44 is required in the yellow phase circuit (marked X) in order to balance the system?

Solution

(a) The reason is to ensure that phases will carry equal currents in order to economically select the correct size cables and switchgear for the installation. It also creates minimal current in the neutral conductor.

(b) With reference to Figure 44 the loading on the red and blue phases are as follows:
Red phase = 84 A + 34 A + 40 A
= 158 A
Blue phase = 75 A + 58 A + 25 A
= 158 A
Therefore yellow phase = 46 A + X A + 48 A = 158 A
Thus X = 158 − 94 = 64 A.

Part II Studies

Theory and Regulations

101 Reproduce the outline of Figure 45 on an overlay and show the conduit routes and cables for the lighting installations taken back to the distribution board.

Solution

See Figure 46.

102 Draw a bar chart of a simple project involving electrical work only.

Solution

The solution to this question is based on Figure 47. The activities for the network are as follows:

0–1	Erection of site hut (½)
1–2	Study of drawings and site (½)
2–3	Start erecting conduit (1)
2–4	Start erecting switchgear (2)
3–6	Continue conduit installation (6)
3–9	Wire conduit system (5)
4–5	Install sub-main cables (3)
4–8	Erect trunking system (6)
5–11	Install disboard cables (8)
6–7	Terminate conduits to machines (1)
7–11	Wire and connect machines (3)
8–10	Wire trunking system (½)
8–12	Install machines (8)
9–14	Connect lighting installation (10)
10–13	Label disboards and machines (1)
11–14	Wire disboards (2)
12–15	Test and commission machines (8)
13–15	Test and commission lighting (3)
14–15	Clear site (½)
15–16	Hand over electrical installation (½)

Note: Figures in brackets are times measured in days.

Figure 48 is a bar chart for the project. The dotted line illustrates those activities taking the longest time to complete the project.

103 Prepare notes on the following topics:
(a) clocking on and off routines
(b) tea- and meal breaks
(c) washing and changing facilities
(d) site meetings

Solution

(a) This method of time-keeping lends itself favourably to employers who have a large workforce, where accurate time-keeping is essential to encourage punctuality. The system often preferred is the punch card method which takes a relatively short time to carry out and provides the

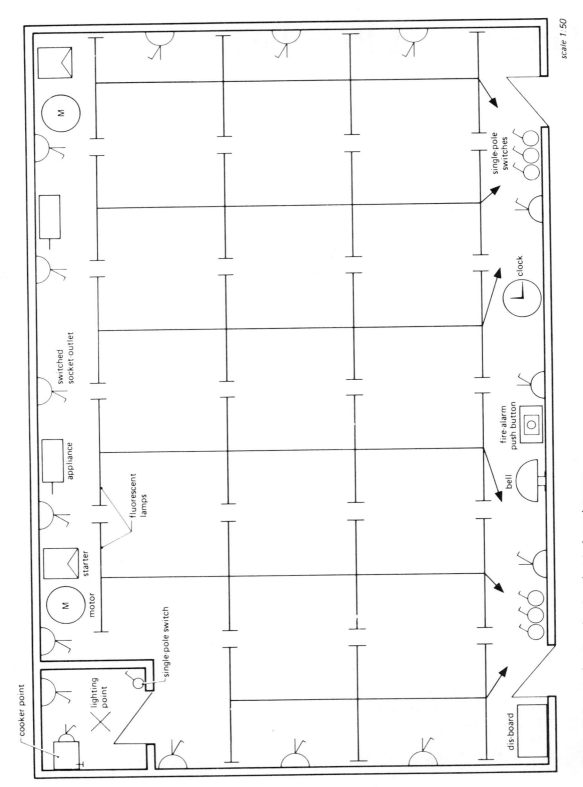

Figure 45 *Layout drawing showing electrical requirements*

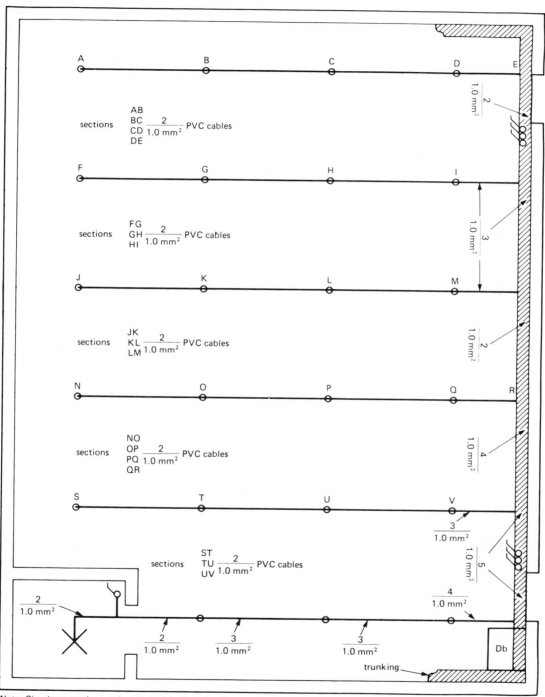

Figure 46

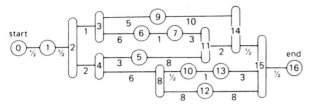

Figure 47

employer with immediate information of employees' time-keeping to work out wages.

In service industries like the electrical contracting industry, the punch card system might be difficult to operate, mainly because of contracts worked away from the office and overtime working. In such circumstances, a good charge-hand or foreman is all that is required to organize a small workforce.

(b) These are taken at set times laid down by the employer. Meal breaks, including washing time, are generally of one hour duration and are unpaid, whereas tea-breaks, which are usually ten/fifteen minute periods in the morning and afternoon, are paid by the employer.

It is in the interest of the employees not to exceed the times granted by their employers as this will only result in lost production time and lost profit for the employer and eventual unemployment for the employees.

(c) These facilities will be found on the premises of the work place. They may be provided by the employer in a site hut or cabin or provided by the main builder, contractor or client as a shared facility. Such facilities will enable the employees to maintain a reasonable standard of hygiene at his/her place of work.

(d) These are usually arranged by the main builder of a large contract to discuss its progress/delay/alterations, etc. The meetings bring together interested parties such as electricians, plumbers and carpenters, etc. Generally it is the charge-hands and foremen of the respective trades who become involved and who have the opportunity to discuss their problems. Such meetings should aim to improve the relations between the different contractors and allow a free flow of work for ensuing weeks until the next meeting.

It is good practice to keep a record of the minutes of each meeting and for them to be agreed and signed before each new meeting commences.

104 What is meant by the following with regard to electrical contracting:
(a) contract
(b) specification
(c) variation order
(d) bills of quantities

Solution

(a) A contract is an agreement which contemplates and creates an obligation.
(b) This is a document which specifies the requirements for carrying out work.
(c) A variation order is an architect's instruction (AI) used for the purpose of carrying out changes or additions to work.
(d) A bill of quantities is a list of all the components of an installation which have been compiled by a quantity surveyor.

105 What is the function of the following bodies:
(a) trade union
(b) Joint Industrial Board (JIB)
(c) Advisory, Conciliation and Arbitration Service (ACAS)

Solution

The answer to this question can be found on page 27 of *Electrical Installation Technology 1* under the heading '*Industrial relations*'. Briefly:
(a) A trade union ensures that its members are provided with a reasonable standard

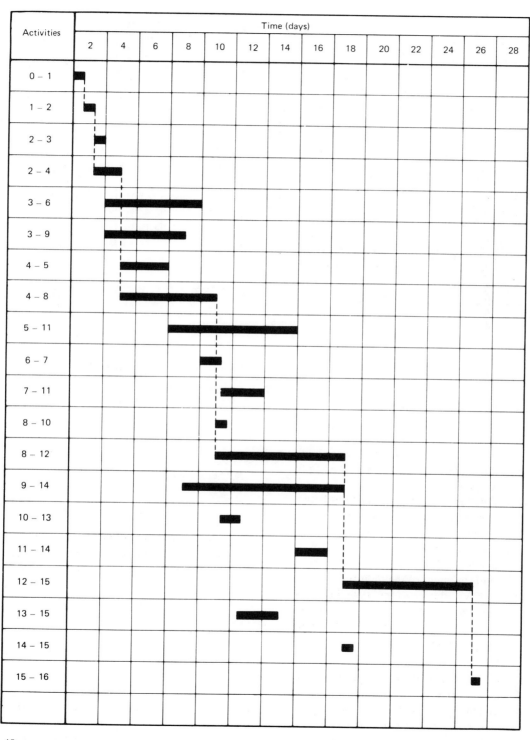

Figure 48

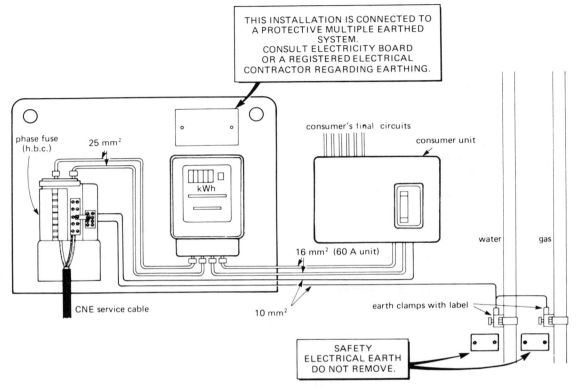

Figure 49 *Motor p.f. correction*

of wages and conditions at work and sees that its members are not victimized by employers.
(b) The JIB regulates the relations between employers and employees within the industry by providing benefits to those persons concerned for the purpose of stimulating and furthering the improvement of the industry.
(c) ACAS arbitrates impartially in decisions between employers and unions where talks on work matters have previously broken down.

106 Draw a diagram of the intake position of a consumer's premises supplied for p.m.e. connection. Show also the concentric service cable joined to the cut-outs, the meter connection, fuseboard connections and bonding arrangements.

Solution

See Figure 49.

107 Make a neat circuit diagram of a 110 V electric drill fed by a double wound transformer with its primary winding connected to a 240 V single phase supply. Show earthing of the transformer and electric drill.

Solution

See Figure 50.

108 (a) Show by means of a diagram how a standard 415 V, 3-phase 4-wire supply to a factory could be used to feed the following loads:

Theory and Regulations

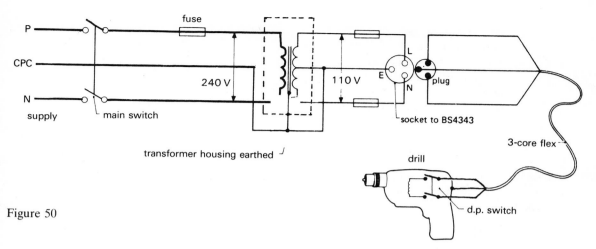

Figure 50

(i) 240 V discharge lighting arranged so as to minimize stroboscopic effect
(ii) one 415 V three-phase motor
(iii) two 415 V single-phase welders

(b) (i) state why balancing of loads is desirable
(ii) state why a neutral is essential on the above three-phase supply

CGLI/II/83

Solution

(a) This part of the question requires diagrams of how the loads are divided over the 3-phase, 4-wire supply. For this see Figure 51.

(b) (i) It should be realized that a balanced system allows equal loading on each phase of the supply. By doing this the correct size switchgear and cables can be chosen. It would be a nonsense to have one phase taking a large load while the other two were taking small loads since selection of a main cable is based on the maximum design current of the system over the three phases. Two lightly loaded phases would be uneconomical resulting in poor utilization.

(ii) It is very difficult to balance a system completely because certain loads are running at different times to other loads. Lighting on one phase may be switched off and the welding plant on and off at

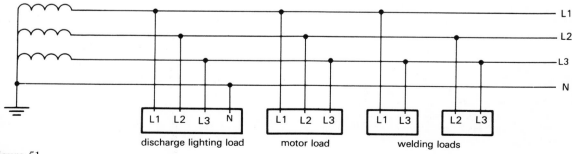

Figure 51

different times. With the exception of the 3-phase motor, the system above is unbalanced and it is the function of the neutral to take back to the supply transformer the unbalanced currents as and when they occur.

109 Determine the maximum permitted earth fault loop impedances for the following final circuits:
(a) A 240 V immersion heater protected by a 15 A semi-enclosed fuse (BS 3036). See Table 41A2(c) of the IEE Wiring Regulations. What current is likely to flow to rupture the fuse?
(b) A 240 V motor protected by a 32 A h.b.c. fuse (BS 88). See Table 41A2(a) of the IEE Wiring Regulations. What current is likely to flow to rupture the fuse?
(c) A 240 V automatic control circuit protected by a 10 A circuit breaker (BS 3871 Type 2). See Table 41A2(e) of the IEE Wiring Regulations. What current is likely to flow to trip the circuit breaker?

Solution

(a) An immersion heater is a fixed piece of equipment and Regulation 413–4 (ii) requires a circuit disconnection within 5 seconds. From Table 41A2(c), the maximum earth fault impedance is 5.6 ohms. At this impedance the fault current would be:

$$I_F = \frac{V_P}{Z_S}$$

where V_P is the phase voltage
Z_S is the loop impedance

Thus $I_F = \frac{240}{5.6} = $ **42.9 A** (approx.)

Note: A semi-enclosed fuse normally ruptures at about twice its current rating. See Figure 11 IEE Wiring Regulations and determine disconnection time.

(b) The motor is regarded as fixed equipment and it will be seen from the appropriate Table that its maximum earth loop impedance is 1.8 ohms. Under these conditions the fault current would be:

$$I_F = \frac{240}{1.8} = \textbf{133.3 A}$$

See Figure 8 of IEE Regulations and check disconnection time.

(c) This circuit is also seen as fixed equipment and from the Table the maximum earth loop impedance is 3.4 ohms. Under these conditions the fault current would be:

$$I_F = \frac{240}{3.4} = \textbf{71 A}$$

See Figure 14 of the IEE Regulations and check disconnection time.

110 Describe the internal layout of a metalclad, six-way consumer unit fitted with m.c.b.s.

Solution

A metal clad consumer unit is a factory-built assembly (see 'Definitions', IEE Wiring Regulations, page 9) and for indoor use, designed to the specification of BS 5486, Part 13 (1979). The two main features of the consumer unit will be miniature circuit breakers and main switch which often takes the form of a double-pole earth leakage circuit breaker of the residual current type (i.e. RCD), rated between 30 A and 100 A.

The miniature circuit breakers are to BS 3871 specification and come in current ratings of 5 A, 10 A, 15 A, 20 A, 30 A, 40 A and 50 A. These devices are arranged in line with each other with a common busbar linking the lower m.c.b. terminals and the outgoing phase connection of the RCD. The RCD also has an outgoing multi-terminal neutral and the device provides protection against current leakage to earth with tripping values commonly at 30 mA, 100 mA and 300 mA.

Theory and Regulations 67

The interior assembly is often mounted on a removable sub-plate to allow the consumer unit to be fixed in position. The consumer unit enclosure will have a conveniently positioned earth bar and ample knockout entry holes for final circuit wiring. All units will be provided with circuit identification labels and instruction information. It is important, when wiring inside the unit, to terminate the conductors correctly, making sure that circuit earth and neutral conductors follow the same wiring order as the phase conductors.

111 Write down a meaning for the following terms:
(a) fuseboard
(b) consumer unit
(c) emergency switch
(d) residual current device
(e) indirect contact

Solution

(a) *fuseboard* An assemblage of parts, containing one or more fuses or circuit breakers arranged to provide final circuits with excess current protection.
(b) *consumer unit* As in (a) above, but with a main switch incorporated for controlling consumer's final circuits.
(c) *emergency switch* A device for rapidly cutting off the supply of electrical energy.
(d) *residual current device* A mechanical switching device intended to cause the opening of contacts when the device reaches a specific operating condition.
(e) *indirect contact* This means contact of a person (or livestock) with exposed conductive parts that have become live owing to a fault.

112 From Figure 45 (see page 60), make a list of all electrical equipment required.

Solution

See Table 1.

Table 1

Qty.	Description	Cat. no.	Price
23	Fluorescent luminaires	Thorn PP40	
15	13 A switched socket outlets	MK 2977 ALM	
1	Distribution board	Wylex HS91	
1	Metalclad cooker control point	MK 5001	
1	1-Gang, 1-way s.p. switch	MK 3591 ALM	
2	3-Gang, 1-way s.p. switch assembly	MK 892 ALM (complete)	
1	Dome bell	Gent 500AC	
1	Fire-alarm point	Gent 1102	
2	Redring fan-heaters	50–811103	
1	Battenholder	MK 1173 WHI	
2	Memdol single-phase a.c. motor switches	MEM 2 SPS + 24 SPH	
2	Brook Crompton Parkinson Motors		
1	Fused connector box for electric clock	MK 993 WHI	
1	Tann Synchronome clock		

113 Show the circuit wiring of a domestic consumer's installation comprising the following circuits:
(a) upstairs and downstairs lighting on separate circuits
(b) immersion heater circuit (night supply)
(c) cooker circuit
(d) two ring circuits
(e) storage heater circuit

Solution

See Figure 52.

114 In the process of calculating the loading on the main cable of a hotel installation, a diversity on cooking appliances was applied. From Table 4B, Appendix 4 of the IEE Wiring Regulations, determine the loading for five cookers each rated at 15 kW, 240 V.

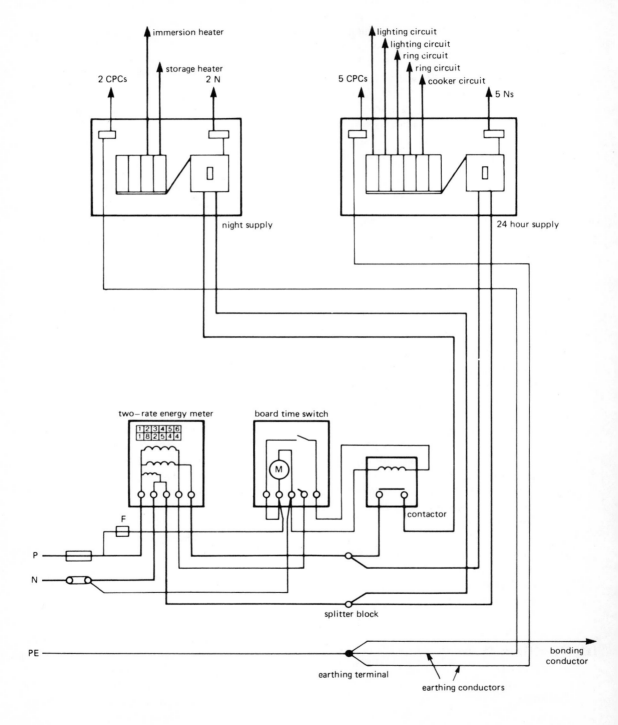

Figure 52

Solution

Current loading of one appliance is:

$$I = \frac{P}{V} = \frac{15\,000}{240} = 62.5 \text{ A}$$

From Table 4B the allowance is 100% f.l. of largest appliance, 80% f.l. of second largest appliance and 60% f.l. of remaining appliances. Thus the assessed current demand is:

$62.5 + (0.8 \times 62.5) + (0.6 \times 62.5 \times 3) =$ **224.5 A**

The hotel would probably be supplied with 3-phases, 4-wire in which case the cookers would be balanced across the phases with other loadings.

115 A 240 V, single 2-phase domestic installation consists of the following equipment:
12 tungsten filament lighting points
10 fluorescent lighting points, each rated at 85 W
2 ring circuits supplying 13 A socket outlets
An immersion heater rated at 3 kW with thermostat control
A 14 kW cooker with a 13 A socket outlet on the control unit.
(a) Calculate the assumed current demand for this installation.
(b) State the nearest standard current rating of the main isolator required.

CGLI/II/88

Solution

(a) Reference should be made to Table 4A and 4B of Appendix 4, the IEE Regulations.
GLS lamps have a current demand of 100 W per lampholder and for a household premises a diversity allowance of 66% applies.

Thus $I = P/V$
$= (1200 \times 0.66)/240$
$= \mathbf{3.3\ A}$

MCF lamps use a multiplier of 1.8 if insufficient information is given of control gear. The same diversity allowance is applicable.

Thus $I = P/V$
$= (10 \times 85 \times 1.8 \times 0.66)/240$
$= \mathbf{4.2\ A}$

The ring final circuits will be based on their overcurrent protective device ratings in accordance with Table 5A of Appendix 5. Each circuit will be assumed at 30 A.
From Table 4B, diversity allowances of 100% and 40% are applicable to the two circuits.

Thus $I = 30 + (0.4 \times 30) = \mathbf{42\ A.}$
For the immersion heater there is no diversity allowance.

Thus $I = P/V = 3{,}000/240 = \mathbf{12.5\ A}$
For the cooker circuit an allowance is applicable based on the formula:
$I = 10 + (0.3 \times R) + 5$ where R is remainder of the rated current.
Thus $I = P/V = 14{,}000/240 = \mathbf{58.33\ A}$
$I = 10 + (0.3 \times 48.33) + 5$
$= \mathbf{29.5\ A}$

For the installation the assumed current demand is

$I = 3.3 + 4.2 + 42 + 12.5 + 29.5$
$= \mathbf{91.5\ A}$

(b) The size of the main isolator required is 100 A.

116 (a) Explain what is meant by the following terms when applied to installation circuits:
(i) diversity
(ii) assumed current demand
(b) In a single-phase, 240 V installation for a small business premises, a circuit is required to feed a distribution board for a number of final circuits. Details of the points of utilization and current-using equipment are shown in the table below.

LIGHTING	12	100 W luminaires (filament lamps)
	16	65 W fluorescent luminaires
HEATING	1	3 kW fixed heater
and POWER	2	1 kW fixed heaters
	2	500 W appliances (office equipment)
MOTORS	1	4.8 kVA input
	1	2.4 kVA input
	1	1.2 kVA input
WATER HEATER	1	5 kW water heater (thermostatically controlled)
SOCKET OUTLETS		Standard circuit arrangement
	1	A2 radial circuit (32 A)
	1	A3 radial circuit (20 A)

Using relevant IEE tables and making allowances for diversity where possible, determine:
(i) the current demand for **each** of the load categories
(ii) the total assumed current demand for the cable feeding the distribution board.

CGLI/II/87

Solution

(a) (i) Diversity is used in installation design in order to determine the maximum current demand of circuits to enable an electrical designer to economically assess the size of cables and switchgear. The term is often expressed as a percentage allowance or diversity factor and it takes into consideration certain circuit and current-using equipment that it is estimated will never be fully utilised. For example, it is quite easy to see that a lighting final circuit in a domestic dwelling might supply numerous outlet points and these would seldom be switched on altogether. A cooking appliance is another example.

(ii) The term 'assumed current demand' is a method of estimating the loading of current-using equipment. A GLS lighting point is assessed on the basis of its current equivalent to the connected load with a minimum of 100 W irrespective of lower wattage lamps being fitted. In the case of a discharge lamp, knowledge of its control gear power factor may not be known and assessment here is based on using a 1.8 multiplying factor. For further details, see Table 4A and Note in Appendix 4, IEE Regulations.

(b) Reference should be made to Table 4B of the IEE Regs.
GLS lamps:
A diversity allowance of 90% is applicable
Design current $I = (P \times 0.9)/V$
$= (12 \times 100 \times 0.9)/240$
$= \mathbf{4.5\ A}$

MCF lamps:
A diversity allowance of 90% is applicable
Design current $I = (P \times 0.9 \times 1.8)/V$
$= (16 \times 65 \times 0.9 \times 1.8)/240 = \mathbf{7.02\ A}$

Heating and power:
A diversity allowance of 100% f.l. of largest appliance plus 75% f.l. of remaining appliances
Design current $I = [(P \times 1)/V + (P \times 0.75)/V]$
$= [3000/240 + (3000 \times 0.75)/240]$
$= 12.5 + 9.375$
$= \mathbf{21.875\ A}$

Motors:
A diversity allowance of 100% f.l. of largest motor plus 80% f.l. of second largest motor plus 60% f.l. of remaining motors.

Design current $I = [(P \times 1)/V$
$+ (P \times 0.8)/V$
$+ (P \times 0.6)/V$
$= [4800/240 + (2400 \times 0.8)/240 + (1200 \times 0.6)/$
$= 20 A + 8 A + 3 A =$ **31 A**

Water heaters:
There is no diversity allowance.
Design current $I = P/V$
$= 5000/240 =$ **20.83 A**

Socket outlets:
A diversity of 100% of current demand of largest circuit plus 50% of current demand of every other circuit.
Design current $I = 32 + (0.5 \times 20) =$ **42 A**

The total assumed current demand is as follows:
$I = 4.5 + 7.02 + 21.875$
$+ 31 + 20.38 + 42$
$=$ **127 A** (approx.)

117 Describe how you would fix a large metal busbar chamber to a brick wall using rawlbolts.

Solution

A typical large busbar chamber is one of 500 A rating and measuring some 527 × 1 384 × 175 mm (HWD). It will have switchgear and fusegear mounted above and below it and room must be allowed for these items of apparatus. Having established its correct height, the busbar chamber is offered up to the wall and marked accordingly using a spirit level to obtain horizontal and vertical alignment. Also, check the wall markings against the brickwork line.

The two most common methods of drilling the wall to fix rawlbolts are (a) using a stardrill and (b) using a tipped masonry drill. The latter method is preferred and includes the use of a hammer drill and rotary impact drilling machine. Whatever method is used, eye protection should be worn. The correct drill bit must be chosen for the rawlbolt size outer diameter and often bolt projecting rawlbolts are chosen for this purpose. The base of these bolts is shaped to form an expander wedge and when the nut is tightened against the outer fixing, the expander wedge is drawn through the rawlbolt shell to provide a very tight fixing.

Only the thread of the rawlbolt should be showing for each hole and the busbar chamber mounted on these and erected in position, applying the nut and washer to secure the fixing of each rawlbolt.

118 Give a meaning for the following terms:
(a) accessory
(b) ambient temperature
(c) bonded
(d) extraneous conductive part
(e) prospective short circuit current

Solution

(a) An accessory can be any device connected with wiring and current-using appliances such as plugs and socket outlets but excludes electrical equipment such as a luminaire or motor.
(b) This is the temperature of the air or other medium where equipment is to be used.
(c) This is a term associated with items of metalwork for connection electrically to ensure a common potential exists. A bonding conductor is used for this purpose.
(d) This is a conductive part which is liable to transmit a potential including earth potential but not itself forming part of the electrical installation.

(e) This is the term given to explain the magnitude of the fault current that may be likely to flow under extreme short circuit conditions.

119 What is the difference between a PME system (TN–C–S system) and an SNE system (TN–S system)?

Solution

The abbreviation PME means *protective multiple earthing* whereas SNE means *separate neutral and protective conductors*. In the former case, the method is commonly used where the source of energy incorporates multiple earthing of the neutral – the letter *C* in TN–C–S means that the neutral and protective conductors are common in a single conductor called a *PEN* conductor or combined neutral earth conductor (CNE). The PEN conductor is only taken as far as a consumer's supply terminals and from this point on in the premises, separate neutral and earth conductors are taken. The letter *S* in TN–C–S indicates this.

With regard to the SNE system, this system involves both a separate neutral and separate protective conductor run from the supply source to separate terminals at each consumer's premises. In practice, the supply cable sheath/armouring provides the protective function and the neutral conductor is incorporated in the cable along with the phase conductors.

120 (a) What are the requirements of the IEE Wiring Regulations regarding BS 1363 socket outlets on a ring final circuit?
(b) Draw a circuit diagram of a permanently connected appliance fed from a fused spur taken from a ring circuit.

Solution

(a) There are a number of requirements in the IEE Regulations – see Appendix 5.

(i) In domestic premises the floor area served must not exceed 100 m².
(ii) Consideration must be given to the loading of the circuit, particularly the loading in kitchens which may require a separate circuit.
(iii) An unlimited number of socket outlets may be provided. A twin or multiple socket outlet is regarded as one socket outlet.
(iv) The protective device at the origin of the circuit must be rated at 30–32 A and size of PVC insulated copper cables is 2.5 mm², PVC insulated copperclad aluminium is 4 mm² and MIMS copper cables is 1.5 mm².
(v) Immersion heaters and the like in excess of 15 litres capacity should be supplied from their own circuits.
(vi) The number of non-fused spurs must not exceed the number of socket outlets and stationary equipment connected to the ring.
(vii) A non-fused spur must not feed more than one single or twin socket outlet, or more than one fixed appliance.
(viii) Permanently connected equipment must be fed from a fused spur rated at not more than 13 A and controlled by a switch or circuit breaker of rating not exceeding 16 A.

(b) See Figure 53.

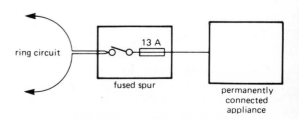

Figure 53 *Ring final circuit connections*

121 A 95 mm² two-core armoured PVC insulated cable (copper conductors), clipped direct to a surface, forms a two-wire radial distributor loaded at three points. From the supply end at 250 V, the first load point is 100 A and is at a distance of 50 m. The second load point is 80 A and is a further 25 m. The third load point is 45 A and is 40 m away from the second load point. Using Appendix 9, Table 9D4 of the IEE Regulations, calculate the voltage at the three load points.

Solution

Voltage drop is found by the formula
V = Length (m) × design current (I_B) × millivolt drop/ampere/metre (mV/A/m).
To the first load point, the cable carries a current of 225 A.
To the second load point, the cable carries a current of 125 A.
To the third load point, the cable carries a current of 45 A.
From Table 9D4 (contd.), the mV/A/m is 0.5. Since each length is known, then:

(i) Voltage drop at first load point

$V = 50 \times 225 \times 0.5 \times 10^{-3}$ = **5.6 V**

Voltage at first load point

250 V − 5.6 = **244.4 V**

(ii) Voltage drop at second load point

$V = 25 \times 125 \times 0.5 \times 10^{-3}$ = **1.56 V**

Voltage at second load point

244.4 − 1.56 = **242.8 V**

(iii) Voltage drop at third load point

$V = 40 \times 45 \times 0.5 \times 10^{-3}$ = **0.9 V**

Voltage at third load point

242.8 − 0.9 = **241.9 V.**

122 (a) State the required disconnection time for final circuits feeding:
  (i) 13 A socket outlets **not** installed in accordance with Appendix 7
  (ii) fixed equipment.
 (b) Determine the maximum permitted earth fault loop impedance (Z_S) for:
  (i) a 240 V electric cooker control unit (without socket outlet) protected by a 30 A fuse to BS 1361
  (ii) a ring final circuit supplying 13 A socket outlets and protected by a 30 A fuse to BS 3036
  (iii) a lighting final circuit protected by a 5 A Type 1 m.c.b. to BS 3871.

Solution

(a) (i) 0.4 seconds [see Reg. 413–4(i)]
   (ii) 5 seconds [see Reg. 413–4(ii)]
(b) (i) 2.0 ohm [see IEE Regs. Table 41A2(b)]
   (ii) 1.1 ohm [see IEE Regs. Table 41A1(c)]
   (iii) 12 ohm [see IEE Regs. Table 41A2(d)]

123 Select an appropriate cable size from the Tables of the IEE Wiring Regulations for a circuit having the following details:
 (a) PVC single-core cables in conduit on a wall
 (b) load – single-phase 240 V taking 30 A
 (c) ambient temperature 30 °C
 (d) length of run 43 m
 (e) overcurrent protection is by m.c.b.
 (f) conduit already contains five single-phase circuits

Solution

The procedure is as follows:
 (i) Design current of circuit = 30 A.
 (ii) Protective m.c.b. rated at 30 A (see Table 41A2 of the IEE Wiring Regulations).
 (iii) Follow procedure in Appendix 9, (6.1.2). Correction factor for ambient temperature does not apply. Correction factor for grouping is 0.57 (see Table 9B).

(iv) Select table from Table 9D1 which can carry a current equal to or greater than:

$$\frac{30}{0.57} = 52.6 \text{ A}$$

From Table 9D1 columns 1 and 4, cable chosen is 10 mm² with current carrying capacity of 57 A.

(v) Check volt drop to see if it is less than the permissible volt drop allowed, i.e. 2.5% of 240 V = 6 V.

Volt drop = 43 × 30 × 0.0044 = 5.67 V

The above cable is suitable for the load.

Note: The question assumes the conduit does not contain any circuit protective conductors.

124 Determine the size of cable trunking for the following conductor numbers and sizes:
(a) fifteen 1.5 mm²
(b) twenty 2.5 mm²
(c) six 4.0 mm²
Assume all are stranded PVC single-core cables. Refer to Appendix 12 of the IEE Wiring Regulations.

Solution

This is a straightforward question and should present no difficulty. Reference should be made to Table 12E and 12F. In Table 12E, 1.5 mm² has a factor of 8.1; 2.5 mm² has a factor of 11.4, and 4.0 mm² a factor of 15.2.
Thus:
total cable factor = (8.1 × 15) + (11.4 × 20) + (15.2 × 6)
= **440.7**

Referring to Table 12F, the nearest size trunking is 75 × 25 mm having a factor of 738.

125 Describe the advantages and disadvantages of the following systems:
(a) catenary wiring
(b) earth concentric wiring
(c) pre-fabricated wiring
(d) flexible conduit
(e) underfloor duct system

Solution

(a) This system is often used as an alternative to burying a wiring system in the ground, which could prove costly. It is ideal for temporary supplies and supplies needed for outbuildings as found in agricultural and horticultural installations. The wire used is galvanized steel wire which is strained tight. From this the wiring system cable, e.g. rubber or PVC sheathed cable, is taped or suspended by hangers. Some systems have integral cable and catenary wire. The IEE Regulations restrict the height of aerial cables incorporating catenary wire to 3.5 m above ground where vehicles are inaccessible and 5.2 m where they are accessible, apart from road crossings where the minimum height is 5.8 m. It should be noted that the 3.5 m height mentioned is not applicable to agricultural premises.

In terms of disadvantages, the catenary wiring system is restricted in use and has not the flexibility of use of other wiring systems found today, such as conduit and trunking systems where circuits can be easily altered. The catenary system does not provide a high degree of mechanical protection.

(b) In this system, the immediate advantage is found by the sheath being used as a return conductor. Basically, the wiring system used is MIMS cable since its outer sheath is copper, which provides an ideal PEN conductor. Sealing pots which contain an earthing tail are used for this purpose. While the system has the advantage of not requiring any neutral conductors contained within the cable itself, its use as a wiring system is somewhat restricted to installations not connected directly to the public supply.

Furthermore, where MIMS is used as the wiring system, the sheath must not have a c.s.a. of less than 4 mm^2 and its resistance must never be more than any of its internal conductors. Other conditions for its use are given in the IEE Wiring Regulations, Regs. 546–1 to 546–8.

(c) This system has the advantages of reduced cost of site installation time since it is prepared off-site. It is often supplied pre-wired, depending on the type of system, with cables cut to length ready for termination. In practice, the system lends itself to duplicate installations, such as domestic dwellings, and it is important that manufacturers' site information is adhered to during installation. Multi-bore and Octoflex are typical systems used today.

The disadvantages of the prefabricated systems are their restrictions in general use, flexibility of installation practice and site alterations.

(d) Flexible conduit systems have a somewhat restricted use as a wiring system, their main advantage being where electrical equipment and apparatus requires moving or is vibrating, such as the wiring to a motor. The system should preferably be waterproof and if of metallic design its enclosure should not be relied upon as a protective conductor.

(e) This system has particular use in commercial premises, particularly large office blocks with requirements for numerous business machines, electric typewriters, duplicators, photocopiers, etc., spread about the building. While the system has numerous advantages in the premises described, it is again somewhat restricted, since once installed it cannot be modified without a great deal of expense and inconvenience. Careful planning at the design stages of the building are essential as well as at the installation stage to avoid damage to the system before it is covered *in situ*.

126 What are the IEE Wiring Regulations concerning:
(a) space factor for cables drawn into ducts
(b) single-core, PVC cables drawn through metal conduit
(c) cables unsuitable for a.c.
(d) maximum operating temperature of PVC compound cables contained within luminaires?

Solution

(a) Regulation 529–7 points out that the number of cables laid in the enclosure of a wiring system shall be such that no damage is caused to the cables or its enclosure. The IEE Regulations provide no data on percentage limitations and one should seek information from duct manufacturers.

(b) The requirement for single-core cables is spelt out in Reg. 521–8, and the enclosure of metal conduit around a single-core conductor creates eddy currents (see Figure 54) which cause the conduit to become hot.

(c) See Reg. 521–8 again – the reason is given in (b) above.

(d) See section 523, IEE Wiring Regulations, 'ambient temperature', Reg. 523–3 and 422–4.

Note: Heat resisting flexible cords are recommended between a ceiling rose and lampholder such as silicone rubber or glass-fibre, particularly in luminaires using tungsten lamps.

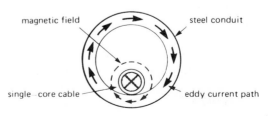

Figure 54 *Circulation of eddy currents*

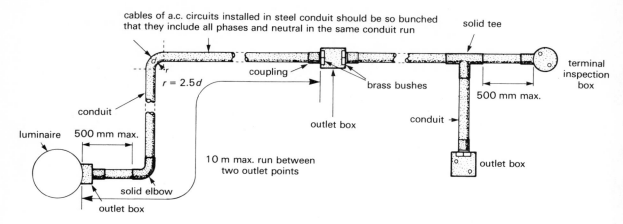

Figure 55 *Conduit requirements*

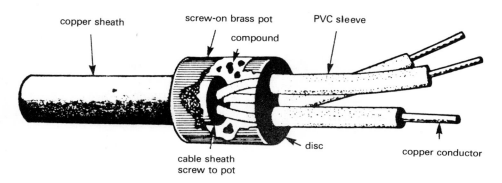

Figure 56 *MI cable termination*

127 Make a neat sketch showing how solid conduit elbows and tees can be used, complying with Regulation 529–4 of the IEE Wiring Regulations.

Solution

See Figure 55.

128 Show an exploded view of a 3-core MIMS screw-on seal termination. Label all parts.

Solution

See Figure 56.

129 Draw a circuit diagram of the lighting control switching illustrated in Figure 57.

Solution

See Figure 58.

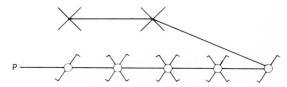

Figure 57

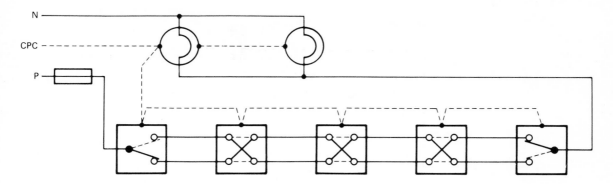

Figure 58

130 With reference to Appendix 12 of the IEE Wiring Regulations, calculate suitable trunking sizes for the following installed cables:
  (a) nine 10 mm² single-core PVC insulated cables, each having an overall diameter of 6.2 mm
  (b) four 16 mm² single-core PVC insulated cables, each having an overall diameter of 7.3 mm

Solution

(a) First, find the c.s.a. of one 10 mm² cable, using:
$$A = \frac{\pi \times d^2}{4}$$
$$= \frac{3.142 \times 6.2 \times 6.2}{4}$$
$$= \frac{120.778}{4}$$
$$= \mathbf{30.19\ mm^2}$$

The overall c.s.a. of the cables is $9 \times 30.19 = \mathbf{271.71\ mm^2}$.

Make reference to Appendix 12 of the IEE Wiring Regulations, Table 12E. If the factor of 36.3 was chosen for each 10 mm² then the overall c.s.a would be $9 \times 36.3 = \mathbf{326.7\ mm^2}$.

However, since specific information is given in the question, then a suitable trunking size from Table 12F is 75 mm × 25 mm having a factor of 738 which satisfies both calculations.

(b) The overall c.s.a. of this cable is:
$$A = \frac{\pi \times d^2}{4}$$
$$= \frac{3.142 \times 7.3 \times 7.3}{4}$$
$$= 41.86\ mm^2$$

For four cables it is $4 \times 41.86 = 167.44\ mm^2$. From the note at the foot of the Tables, use has to be made of the space factor of 45%. Thus:
$$A = 167.44 \times \frac{100}{45} = \mathbf{372\ mm^2}$$

Again the 75 mm × 25 mm trunking will be found suitable.

131 Show by means of a diagram the layout of a single-phase distribution system supplying outlet units for use on a construction site.

Solution

See Figure 59.

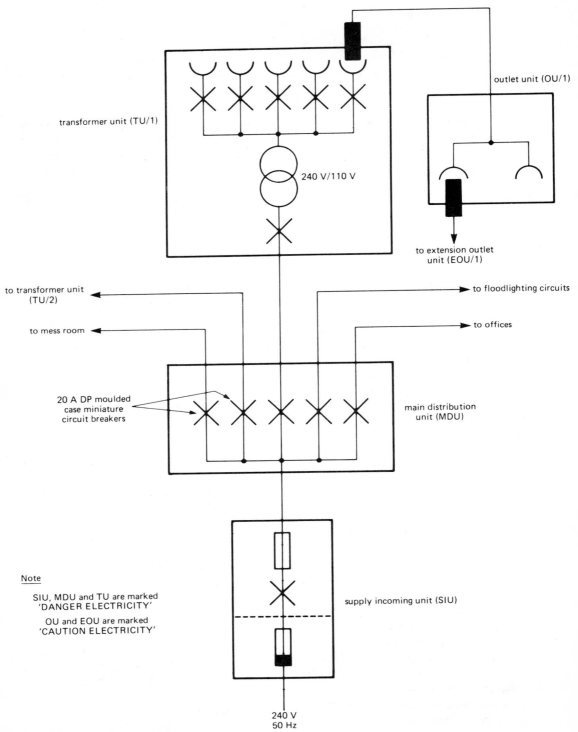

Figure 59 *Construction site supplies*

132 What are the special requirements in the IEE Wiring Regulations for (a) flexible cables and flexible cords, and (b) heating wires and cables?

Solution

(a) Flexible cables and cords are dealt with in Chapter 52 of the IEE Regulations (see Tables 9H2 and 9H3). Generally speaking, flexible cables have sizes ranging from 4.0 mm² to 630 mm², whereas flexible cords range from 0.5 mm² to 4.0 mm².

While Regulation 521–5 (concerning the cables used at low voltage) provides examples of the types of cable, it does not concern itself with flexible cords forming part of portable appliances or luminaires or to those cables for combined power and telecommunications wiring, these latter cables being the subject of Regulation 525–8. This regulation is one of twelve requirements concerned with prevention of mutual detrimental influence. Briefly, circuits are split up into categories of circuits, C1 being a circuit operating at low voltage fed from a mains supply system (it excludes a fire alarm circuit or emergency lighting circuit), C2 being any circuit for telecommunications, such as radio, intruder alarm, data system, etc. (it again excludes a fire alarm and emergency lighting circuit), and C3 being a fire alarm circuit and emergency lighting circuit.

It is Regulation 525–8 which relates to the way the flexible cables and flexible cords should be used with each other or separated from each other, as the case may be, in terms of their category identification. Figure 60 illustrates this arrangement of segregation of circuits.

(b) These regulations are dealt with in Chapter 55 of the IEE Wiring Regulations under the sub-heading: 'Conductors and cables for soil, road and

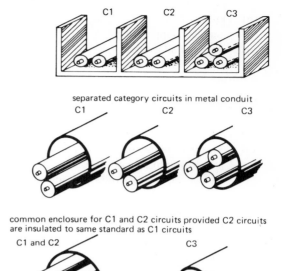

Figure 60 *Category circuits*

floor warming'. Regulation 554–31 requires cables to be enclosed in material having class P ignitability characteristics, specified in BS 467: Part 5. They should also be protected from mechanical damage (see Reg. 554–32) and be constructed of materials that are resistant to damage from dampness and/or corrosion. In Regulation 554–33 the concern for heating cables laid below surfaces is that they should be completely embedded in the substance they intend to heat as well as not suffer damage by any natural movement they may have or from the substance which surrounds them. Reference should be made to Table 55D with regard to maximum operating temperatures for floor warming cables (see also Reg. 554–34).

133 What is meant by:
(a) protective conductor
(b) growth factor
(c) ambient temperature
(d) diversity allowance

Solution

The answer to these terms can be found in *Electrical Installation Technology 1* and the IEE Wiring Regulations.
(a) This is a conductor used to protect against the likelihood of electric shock. It connects together exposed conductive parts, extraneous conductive parts, the main earthing terminal, earth electrode and earthed point of source.
(b) This is a factor which considers the likelihood of future loadings on switchgear and cables.
(c) This is the temperature of the air or other medium where the equipment is to be used.
(d) This is a ratio of minimum actual loading and installed loading based on the assumption that all equipment in an installation will not be in full use at any one time.

134 What are the advantages and disadvantages of the following systems:
(a) ring main system
(b) rising main system
(c) TN–C–S system

Solution

(a) The most important advantage of this system is that load points are fed two ways making full use of conductor c.s.a as a result of diversity applied to the load points. A further advantage is that breakdowns on the system are easier to trace while still retaining continuity of supply elsewhere as well as isolation of part of the system for maintenance purposes. Volt drop problems are also reduced.

There are few disadvantages with a ring main but one cannot ignore the increased cost of cable and duplication of switchgear for isolation purposes. The system is also limited: its greatest use is on primary supply distribution systems.

(b) This is another form of main supply system which lends itself to multi-storey type buildings where lateral supplies are tapped off to feed floors. Systems are available up to 2 000 A using high conductivity copper busbars; this rating and flexibility of supply are seen as its greatest advantages.
Expense may be seen as its greatest disadvantage and this may limit its choice to the types of building mentioned.

(c) The main advantage with this supply system is the reduced cost to the supply authority in terms of cable cores and termination methods on distributor cables. The system is known as p.m.e. where the neutral and protective faults are converted into phase to earth faults.
The main disadvantage of the system is that if the PEN conductor becomes an open circuit, it results in a possible danger to consumers.

135 Determine the maximum disconnection time for the following cables:
(a) 50 mm^2 twin armoured cable with 90° thermosetting insulation and aluminium conductors
(b) 50 mm^2 mineral-insulated cable with copper conductors
(c) 50 mm^2 twin PVC insulated cable with copper conductors

Assume a prospective fault current or effective short circuit current of 4 000 A.

Solution

Reference should be made to the adiabatic equation on page 37 of the IEE Wiring Regulations. This equation is used when the cross-sectional area of conductors is 10 mm^2 or more and for short circuits of duration up to 5 seconds.

(a) $t = \dfrac{k^2 S^2}{I^2}$

$= \dfrac{94^2 \times 50^2}{4\,000^2} = \mathbf{1.38\ s}$

(b) $t = \dfrac{135^2 \times 50^2}{4\,000^2} = \mathbf{2.84\ s}$

(c) $t = \dfrac{115^2 \times 50^2}{4\,000^2} = \mathbf{2.06\ s}$

136 (a) Explain briefly why temporary electrical installations on building sites should be designed to at least the same standard as permanent installations.
(b) For a temporary installation on a building site state the main requirements of the various regulations with regard to:
  (i) type of switches to be used
  (ii) type of plugs and sockets
  (iii) use of overhead cables
  (iv) frequency of testing
(c) State *one* example in *each* case of plant or equipment connected to the following supply voltages:
  (i) 25 V single phase
  (ii) 110 V single phase
  (iii) 240 V single phase
  (iv) 415 V three phase

CGLI/11/83

Solution

(a) In view of the nature of work and hazardous conditions on a building site, a temporary installation is much more vulnerable to misuse than a permanently connected installation. The system may be frequently changed and modified as work progresses and it is important to have a competent person in charge of it. The equipment and wiring systems used should therefore be inspected and tested periodically.

(b) (i) It is recommended that switches controlling single-phase supplies operating up to and including 110 V should be of the double-pole type with circuit protection provided in each live conductor. See BS CP 1017: 1969.
(ii) Socket outlet and plugs should be designed to the standards of BS 4343 and should be either mechanically or electrically interlocked to ensure that the supply to the contact tubes of the accessory is isolated when the plug is withdrawn. Accessories are available for single and three-phase supplies with discrimination between different voltages by colour coding and the positioning of the earth contact in relation to a keyway (see examples in Figure 61).

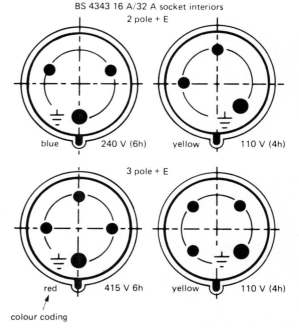

Figure 61

(iii) The use of overhead cables on a building site is not recommended but where it is unavoidable the minimum height of span above ground at road crossings is 5.8 m. In areas where mobile plant is prohibited, cables may be fixed at any height above 5.2 m. Such cables should be bound with tapes, yellow and black in accordance with BS 2929. See BS CP 1017: 1969.

(iv) Temporary installations should be inspected and tested at regular three-month intervals. It is important for checks to be carried out on cable leads and portable tools and proper records kept of all site wiring and equipment.

(c)
(i) Use of 25 V supply is suitable for damp and wet areas.
(ii) Use of 110 V supply is suitable for portable tools.
(iii) Use of 240 V supply is suitable for floodlighting.
(iv) Use of 415 V supply is suitable for large equipment, e.g. crane.

137 What are the IEE Wiring Regulations (15th Edition) with regard to:
(a) provision of a fireman's emergency switch controlling a high voltage neon sign
(b) wiring of capacitors
(c) luminaires
(d) ceiling roses

Solution

(a) Reference should be made to Regulations 476–12 and 476–13. The former regulation deals with unattended interior installations and exterior installations, both operating above low voltage. The latter regulation concerns the provision of the switch: in exterior installations it should be adjacent to the discharge sign, or a notice indicating the position of the switch should be adjacent to the sign. For interior installations, the switch has to be located in the main entrance or position agreed with by the local fire authority. In both cases, the switch height must not exceed 2.75 m from the ground and where more than one switch is installed on any one building, each switch has to be clearly labelled to indicate what it controls. The fire authority needs to be notified of this.

(b) Regulation 461–4 states that means shall be provided for the discharge of capacitive electrical energy. The notes in Regs. 512–1 and 512–2 refer to the adequacy of switches and circuit breakers of capacitive equipment, and Regulation 554–5 states that ancillary equipment for discharge lighting installations, such as capacitors, shall be either totally enclosed in a substantially earthed metal container or placed in a suitably ventilated enclosure to BS 476 Part 5 or be of fire-resistant construction.

(c) There are a number of requirements for luminaires and reference should be made to the index of the IEE Wiring Regulations. Some of these are:
(i) In a bathroom it is recommended to use a totally enclosed luminaire (Reg. 471–38).
(ii) All fixed luminaires are to be placed or guarded to prevent ignition of any materials placed in proximity to them or their lamps – guards which are used must be capable of withstanding the heat from the luminaire or lamp (Reg. 422–4).
(iii) Where a flexible cord supports a luminaire, the maximum mass supported by the cord must not exceed 2 kg for a 0.5 mm^2 cord, 3 kg for a 0.75 mm^2 cord and 5 kg for a 1.0 mm^2 cord (Reg. 524–32).

Also see switching off for mechanical maintenance (Reg. 476–8), emergency switching (Reg. 476–13) and other requirements (Regs. 476–17 to 18).

(d) With regard to ceiling roses, see Regulation 412–6 (an exemption from the requirements to open an enclosure); Regulation 533–19 (must not be installed in any circuit operating above 250 V) and Regulation 553–20 (unless specially designed they cannot be used for the attachments of more than one outgoing flexible cord).

138 What is meant by the following terms:
(a) efficacy
(b) luminous flux
(c) illuminance
(d) utilization factor
(e) maintenance factor?

Solution

(a) *Efficacy* is the term expressing the ratio of luminous flux and power consumed of a lamp.
(b) *Luminous flux* is the flow of light sent out by a light source; it is measured in lumens. There are 4 lumens emitted by a point source of 1 candela.
(c) *Illuminance* is the amount of light in lumens falling on unit area, 1lm/m^2.
(d) *Utilization factor* is a term that shows the light reaching the working plane is reduced so that the power of the light source will have to be increased to obtain the desired illuminance. As a coefficient it expresses the ratio of light flux falling on the working plane and total light flux produced by all light sources.
(e) *Maintenance factor* is expressed as a ratio of illumination for a dirty installation to that from the same installation when clean. The factor allows for dust accumulation on luminaires, ageing of light source and deterioration of decor.

139 Draw a circuit diagram of a high pressure sodium vapour discharge lamp and explain the function of its controlgear. Also describe its colour appearance, efficacy and two typical uses.

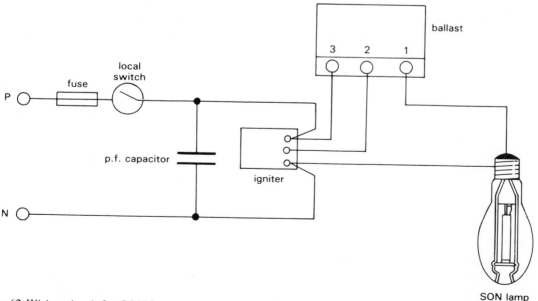

Figure 62 *Wiring circuit for SON lamp*

Solution

See Figure 62 which shows the controlgear and circuitry for a SON lamp. The *ballast* for such a lamp is housed in a container filled with polyester to enable it to withstand a long life. The main function of the ballast is to limit current through the lamp and also provide the lamp with the correct voltage. Some wattage lamps require three ballasts for this requirement.

In order for the lamp to light, a series of high voltage pulses is applied by an *ignitor*. These pulses are of very short duration and because of capacitive attenuation the length of cable between the ignitor and the lamp is limited. For example, the maximum length of cable between lamp and ignitor for a 400 W SON lamp is 17 m. When the lamp is struck the ignitor ceases to function. The lamp takes several minutes to reach full brightness but because the sodium vapour pressure in the arc tube builds up to several atmospheres, the lamp cannot restrike immediately after being switched off – it needs to cool down first.

In terms of colour appearance, SON lamps emit a pleasant golden white light with reasonable colour rendering, so that colours are distinguishable. The efficacy is quite high, for a 400 W lamp with lighting design lumens of 44 000 lm it is 110 lm/w. Two typical uses for the lamp are: (a) sport centres – swimming pools, gymnasiums, etc. (b) floodlighting.

140   (a)   Explain the meaning of *stroboscopic effect*. Why is the condition dangerous in some instances?
         (b)   Show by means of a diagram how twelve fluorescent lamps can be arranged to minimize the above effect.

Solution

(a)   Stroboscopic effect is a phenomenon peculiar to discharge lighting since the a.c. supply passes through zero twice every cycle. This causes the light to flicker at twice the supply frequency and while not noticeable to the eye the effect can make rotating objects look stationary and create a dangerous situation.

In practice, fluorescent tubes produce an 'after-glow' effect which eliminates the problem. Also luminaires can be wired as lead-lag circuits. The method shown in Figure 63 is to arrange the tubes on different phases but this can only be achieved if a 3-phase supply is available.

(b)   See Figure 63.

141   Explain the following:
      (a)   earth fault loop impedance
      (b)   earth electrode resistance
      (c)   prospective short circuit
      (d)   inspection certificate
      (e)   extraneous conductive part

Solution

(a)   For a phase to earth fault, it is the impedance of the circuit starting and ending at the point of earth fault.
(b)   The resistance of a conductor or group of conductors in intimate contact with and providing an electrical connection to earth.
(c)   This is a term given to current which is possible to flow under extreme short circuit conditions.
(d)   This is a certificate given by an electrical contractor or other person responsible for carrying out inspection and testing of an electrical installation.
(e)   This term refers to a conductive part which does not form part of the electrical installation.

Note: See 'Definitions' in the IEE Wiring Regulations, pages 7–12.

142   What are some of the requirements of the IEE Wiring Regulations with regard to measurement of earth electrode resistance?

Theory and Regulations

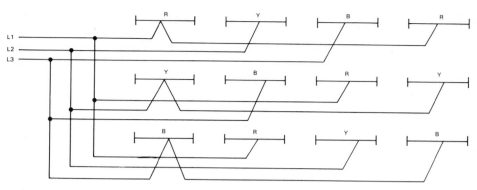

Figure 63 *Balancing discharge lamps on different phases*

Solution

Item 4 of Appendix 15 covers the measurement of earth electrode resistance. Briefly, a test of earth electrode resistance involves the use of either a hand-driven generator tester or mains supply transformer. The former incorporates a rectifier and direct reading ohmmeter while the latter includes a variable resistor to adjust the test current. For tests made at supply frequency, the requirements state that the source of supply must be isolated, hence the use of a double-wound transformer. Also, the electrode under test must be disconnected from its normal earthing conductor. The voltmeter used in this test must have a resistance value in the order of 200 ohms/volt. If not, its resistance will be in parallel with the earth electrode resistance and produce a false voltage reading.

The test requires the average of three readings to be taken where the second auxiliary electrode is moved 6 m close to the test electrode and then 6 m further away from the test electrode. If there is no substantial agreement between the readings then the first auxiliary electrode must be moved further away from the test electrode. The important thing is that these two electrode resistance areas must not overlap.

Note: There are several methods of improving the electrode resistance to a lower value, either driving it deeper into the soil (using extendable type rods) or using chemicals such as common salt and sodium carbonate.

143 Draw a circuit diagram of a residual current device and explain how it operates. What are the IEE Wiring Regulations requirements regarding the testing of e.l.c.b.s?

Solution

A residual current device has already been described in Question 17 and shown in Figure 8.

With regard to testing the devices, reference should be made to the IEE Wiring Regulations, Reg. 613–16 and Item 6 of Appendix 15.

The type shown in Figure 8 is called 'passive' since it does not require electronic amplification for its operating current. The value required to protect users against indirect contact is 30 mA. The test in Item 6 is shown in Figure 116 of *Electrical Installation Technology 1*, page 120 and Regulation requirements are given on pages 00 and 00 of this book.

144 You have been called out to see why a motor is not running. The motor is a three-phase induction motor fed from a distribution board having its own main switch and direct-on-line starter. What procedure would you take to find the fault?

Solution

There may be variations to this solution since the question does not give any indication about

the motor's condition – it may not be running for a number of reasons. Suggestions are:
1. Check rotor is free to rotate – remove belt or coupling.
2. Check for overloading – remove load and check voltage.
3. Test voltage while attempting to start motor – check for volt drop and single-phasing. The motor may make a humming noise.
4. Trace fault back to starter, distribution board, main switch.
5. Check correct fuse sizes, starter overloads, main switch blade action for high resistance contact.
6. Test motor stator windings for any defects.

145 Describe a test to determine the value of earth loop impedance of an earthed concentric wiring system and state what precautions must be taken when carrying out the test.

Solution

An earthed concentric wiring system is a TN–C system in which the PEN conductor is the sheath of the cable – see Figure 3, Appendix 3, IEE Wiring Regulations. A number of the regulations are applicable to PEN conductors – see Section 546 and note below.

In practice there are two basic methods of testing for earth loop impedance (a) using a phase-earth loop tester and (b) using a neutral-earth loop tester. The latter method cannot be used on this wiring system since the earth and neutral are combined in one conductor.

The phase-earth loop test is made by injecting a current (of short duration) of about 20 A from the phase conductor to the PEN conductor. The fault current passes through a known resistor (about 10 ohm) and circulates through the earth fault loop. The value of current can either be measured directly with an ammeter in series with the resistor or using a ballistic instrument calibrated directly in ohms but actually reading the p.d. across the resistor.

As a precaution when making the test, it is advisable to check the PEN conductor first to see if there are any noticeable breaks. Such a break in the system would lead to full supply voltage appearing across the phase and PEN conductors on the test side of the fault.

Note: This system is only permitted where the supply authorities agree. Otherwise it can only be used where the mains are taken through a transformer or converter in order to avoid the supply authority's metallic connection. Further approval of its use can be obtained for a private generating plant.

146 Show diagrammatically how a test for insulation resistance is carried out on a portable electric drill. State ohmic values applicable to the test.

Solution

See Figure 64. Reference should be made to Regulation 613–8 of the IEE Wiring Regulations. Minimum value 0.5 MΩ.

147 Explain the procedure you would take in testing three-phase, six-terminal cage induction motor for:
(a) winding continuity
(b) winding insulation resistance
(c) reversal of direction

Solution

(a) When testing a motor's windings for continuity, all windings must be

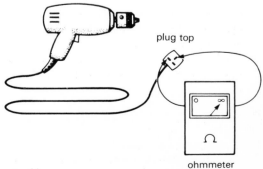

Figure 64

disconnected from each other. Often a multi-range ohmmeter is used and all three windings should give the same ohmic values. It is important for the winding ends to be marked or labelled in case reverse phasing of one winding occurs – thus would make the motor run roughly and it would not take load.

(b) In practice, the insulation resistance to earth of a motor's winding should not be less than 1 megohm. This test is carried out using an insulation resistance tester which has 500 V/1 000 V selection and is capable of measuring high ohmic values. The windings can be tested separately to earth or when connected together and tested to earth.

(c) The first thing to observe is the direction the motor is running. It is important to check that no harm will occur with its connected load after reversal. Disconnect motor from load if in any doubt. Reversal is achieved by changing over any two supply leads on the motor.

148 Explain how you would set about testing a three-phase motor's stator winding for (a) continuity and (b) insulation resistance.

Solution

The testing procedure is the same as that given in Question 147. First, the windings need to be disconnected from the supply and for continuity an ohmmeter is used. If the motor only has three winding ends brought out, test between separate pairs. All three readings should be the same. For insulation resistance, test the windings down to earth. The readings should be above 1 MΩ.

149 A large installation was tested for insulation resistance and found to be 0.3 megohms. Explain the procedure you would take in trying to improve this reading or give reasons why the value is relatively low.

Solution

The ohmic value given in the question is insufficient to meet the requirements of Regulations 613–6 and 613–7 respectively, where for a completed installation the insulation resistance must not be less than 1 MΩ.

As stated in Regulation 613–5, large installations may be divided into groups of outlets each containing not less than 50 outlets. It is quite possible in this instance that there are too many outlets being tested at once, having the effect of lowering the test value since circuits are connected in parallel.

The remedy is to sectionalize the wiring and test on the basis described. There may be a possibility of one or more final circuits giving a higher than normal reading particularly when tested between *live* conductors.

150 What are the IEE Wiring Regulations requirements covering the issuing of:
(a) completion certificate
(b) inspection certificate

Solution

(a) When inspection and testing work of an electrical installation has been carried out, the electrical contractor or installer is required to give a completion certificate to the person ordering the work. The form is set out in Appendix 16 of the IEE Regulations and any defects or omissions revealed are to be made good before the certificate is given. This requirement applies to all work including alterations to existing installations. Here the contractor or installer has to report to the person ordering the work, any defects found in related parts of the existing installation.

(b) An inspection certificate must be attached to every completion certificate. The certificate is shown in Appendix 16. It lists many items of inspection and testing and again, has to be given by the contractor or installer to the person ordering the test. It is recommended that inspections of installations be carried out every five years or less. Any poor test results must be reported on this certificate. Both certificates are required to be signed.

Science and Calculations

151 A workshop is supplied from a 415/240 V three-phase 4-wire a.c. system.
  (a) Explain the advantages of using a three-phase 4-wire system instead of a single-phase system for this type of premises.
  (b) It is desired to measure the power factor of a single-phase motor installed in the workshop using an ammeter, voltmeter and wattmeter.
    (i) Draw a diagram showing the connections for all the instruments.
    (ii) Calculate the power factor of the motor if the readings on the instruments are 6 amperes, 240 volts and 1 008 watts respectively.
                      CGLI/II/80

Solution

(a) There are a number of advantages in using a three-phase system over a single-phase system; they are:
  (i) A higher voltage is obtained which allows greater flexibility with installed plant, e.g. 3-phase motors can be used instead of 1-phase motors, two-phase welding equipment can be used and larger capacity plant can be used.
  (ii) Three-phase motors and other plant as well as lighting can be spread over three-phases to balance the system.
  (iii) Any out-of-balance load currents occurring in the system will flow down the neutral and keep the system balanced.
  (iv) Volt drop problems are reduced.
  (v) Stroboscopic effects can be eliminated using three-phase supplies when the lighting is discharge lighting.

(b) (i) See Figure 65.
    (ii) power factor $= \dfrac{\text{power}}{\text{voltamperes}}$

$$= \frac{1\,008}{240 \times 6}$$

$$= \mathbf{0.7\ lagging}$$

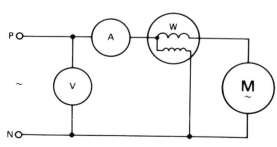

Figure 65 *Method of finding power factor*

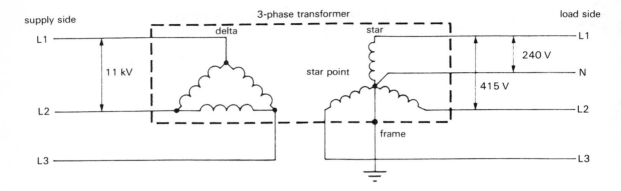

Figure 66 (a) 3-phase, 4-wire system

152 A factory supply is 11 kV, three-phase 3-wire and is fed to a transformer whose output is 415/240 V three-phase.
(a) Draw a circuit diagram of this system.
(b) What is the purpose of earthing the star point of the transformer?
(c) Determine the total kVA load on the transformer if the following three-phase balanced loads were connected:
  (i) 120 kW heating load at unity power factor
  (ii) 240 kVA load at 0.8 power factor lagging.

CGLI/II/80

Solution

(a) See Figure 66.
(b) The purpose of earthing the star point of the transformer is to stabilize the distribution system in order to obtain a neutral connection to allow a lower and safer voltage, phase voltage, to be used for single-phase supplies. Also, any out-of-balance phase currents flowing in the system, as a result of mixed loads, will flow into the neutral conductor and by doing this will keep the system voltage balanced.

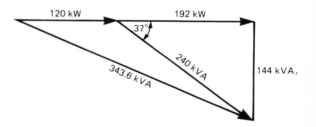

Figure 66 (b) Phasor diagram

(c) (i) For unity power factor conditions, kW = kVA = 120 kW.
   (ii) For 0.8 power factor conditions, kW = 240 × 0.8 = 192 kW.

Adding these two conditions, the active power is 312 kW. The reactive kilovoltamperes of (ii) is:

$Q = \sqrt{S^2 - P^2}$

$= \sqrt{240^2 - 192^2}$

$= 144$ kVAr

Thus the total kVA is

$S = \sqrt{P^2 + Q^2}$

$= \sqrt{312^2 + 144^2}$

$= 343.6$ kVA

See Figure 67 for circuit phasor diagram.

Science and Calculations

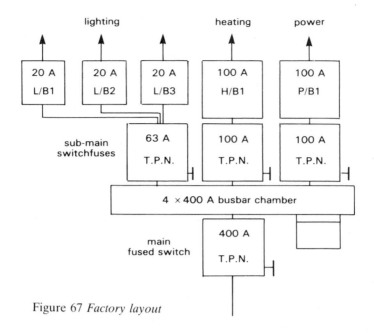

Figure 67 *Factory layout*

153 A factory main switchboard consists of a three-phase, 4-wire feeding a 400 A busbar chamber through a 400 A TPN main fused switch (see Figure 67). On top of the busbar chamber are mounted the following switchfuses:
  (i) A 63 A TPN switchfuse feeding three 20 A SPN lighting distribution fuseboards.
  (ii) A 100 A TPN switchfuse feeding a 100 A TPN heating distribution fuseboard.
  (iii) A 100 A TPN switchfuse feeding a 100 A TPN power distribution fuseboard.
  (a) Draw a diagram showing how interconnectors are made onto the busbar chamber from the main switch.
  (b) Show by means of a line diagram how the fuses are arranged to provide discrimination, assuming the lighting fuseboard fuses are 10 A, heating fuseboard fuses are 15 A and power fuseboard fuses are 30 A.
Note: All switch fuses contain BS 88 Part 2 fuselinks and the fuseboards contain Type 2 and Type 3 miniature circuit breakers.
  (c) Explain the meaning of the term 'fuse discrimination' and briefly state how it is likely to be achieved.

Solution

(a) See Figure 68.

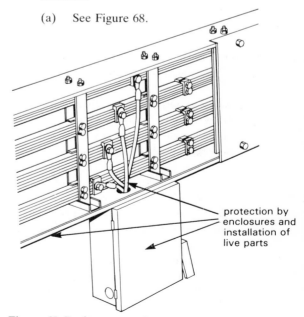

Figure 68 *Busbar connections*

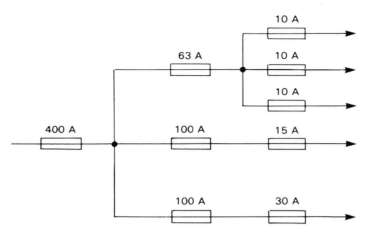

Figure 69 *Fuse discrimination*

(b) See Figure 69.
(c) Overcurrent protective devices in an installation should be so graded that when a fault occurs the device nearest the fault, from the origin of supply, operates first leaving other healthy circuit fuses intact. Consideration should be given to any form of 'back-up' protection required and reference should be made to manufacturers' protective devices to compare time–current characteristics.

154 A resistive heating load of 18 kW consists of three separate elements, each rated at 240 V, 6 kW. The ends of the heating elements are wired into a six-terminal connection block.
(a) Draw a circuit diagram of this arrangement, showing how the six terminals should be interconnected when fed from:
 (i) a 415 V three-phase and neutral supply
 (ii) a 240 V single-phase supply.
(b) For **each** of the above types of supply
 (i) calculate the current in the phase and neutral conductors
 (ii) state the power in the heating load if the neutral conductor becomes disconnected, giving reasons.

CGLI/II/1987

Solution

(a) See Figure 70.
(b) (i) For the three-phase supply, each phase will take a current of
$I = P/V = 6{,}000/240 = $ **25 A**
Since the system is balanced, no current will flow in the neutral conductor.
For the single-phase supply, each resistive heater is in parallel and the phase and neutral current is
$I = P/V = 18{,}000/240 = $ **75 A**
(ii) If the neutral became disconnected in each case, then, for the three-phase supply there would be no effect and the heaters would still operate consuming 18 kW. For the single-phase supply, the neutral carries the return current and the circuit would be unable to operate, consuming no power.

Note: If one element failed to operate when connected to the three-phase supply, then the circuit would become unbalanced and current would flow in the neutral.

155 Figure 71 shows the layout of a 240 V/30 A ring circuit with loads being taken off at points B, C, and D. Given the resistance of each section to be: AB = 0.09 Ω, BC = 0.07 Ω, CD = 0.06 Ω and DA = 0.08 Ω, find:

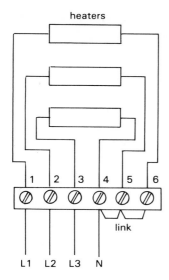

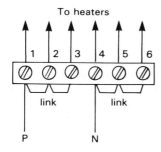

Figure 70 *Circuit connections*

(a) the current in each section of the ring and its direction
(b) the voltages at each load point.

Solution

First, let the current in each section be as follows:
 between AB it is I
 between BC it is $I-10$
 between CD it is $I-23$
 between DA it is $I-26$

Second, since the circuit is a ring circuit the sum of the p.d.s. is zero.

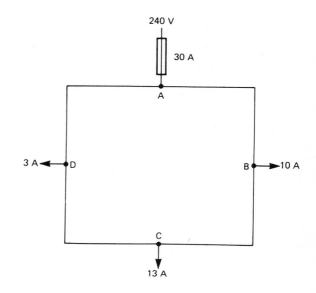

Figure 71 *Ring final circuit*

Thus:
$$0.09I + 0.07(I - 10) + 0.06(I - 23) + 0.08(I - 26) = 0$$
Then:
$$0.09I + 0.07I + 0.06I + 0.08I = 0.7 + 1.38 + 2.08$$
$$0.3I = 4.16$$
$$I = 13.87 \text{ A}$$

(a) Current in each section is as follows:

From AB $I =$ **13.87 A**
From BC $13.87 - 10 =$ **3.87 A**
From CD $13.87 - 23 =$ **−9.13 A**
  (i.e. it flows from D to C)
From DA $13.87 - 26 =$ **−12.13 A**
  (i.e. it flows from A to D)

(b) The voltage at B is:
$$240 - (13.87 \times 0.09) = \textbf{238.75 V}$$
The voltage at C is:
$$238.75 - (3.87 \times 0.07) = \textbf{238.48 V}$$
The voltage at D is:
$$240 - (12.13 \times 0.08) = \textbf{239 V}$$

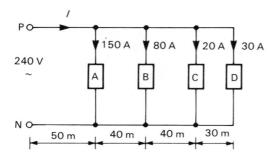

Figure 72 *Radial distribution*

156 With reference to Figure 72, determine:
(a) the supply current
(b) the volt drop across each load A, B, C and D

Assume the cable used has a resistance of 0.1 ohm/1 000 metre.

Solution

(a) The supply current is the addition of the branch currents, i.e.:

$$I = I_A + I_B + I_C + I_D$$
$$= 150 + 80 + 20 + 30$$
$$= 280 \text{ A}$$

(b) Since volt drop $V = I \times R$, then volt drop in cables from origin to load A is:

$$V = 280 \times \left(2 \times 50 \times \frac{0.1}{1\ 000}\right)$$
$$= 280 \times 0.01$$
$$= 2.8 \text{ V}$$

volt drop in cables between loads A and B is:

$$V = 130 \times \left(2 \times 40 \times \frac{0.1}{1\ 000}\right)$$
$$= 130 \times 0.008$$
$$= 1.04 \text{ V}$$

volt drop in cables between loads B and C is:

$$V = 50 \times \left(2 \times 40 \times \frac{0.1}{1\ 000}\right)$$
$$= 50 \times 0.008$$
$$= 0.4 \text{ V}$$

volt drop in cables between loads C and D is:

$$V = 30 \times \left(2 \times 30 \times \frac{0.1}{1\ 000}\right)$$
$$= 30 \times 0.006$$
$$= 0.18 \text{ V}$$

Note: It should be pointed out that the terminal p.d. for load A is 240 − 2.8 = 237.2 V; for load B it is 237.2 − 1.04 = 236.16 V; for load C it is 236.16 − 0.4 = 235.76 V; and for load D it is 235.76 − 0.18 = 235.58 V. At load D the permissible minimum terminal voltage is 234 V (i.e. a 6 V maximum volt drop allowed from a 240 V supply — see IEE Wiring Regulations, Reg. 522.8)

157 In a 415 V/50 Hz, 3-phase, 4-wire supply system, the following loads are connected:
(a) one 50 Ω non-inductive load resistor across the red phase
(b) one 1 000 W load having a p.f. of 0.8 lagging across the yellow phase
(c) one capacitive and resistive load of 20 Ω reactance and 40 Ω resistance in series with each other across the blue phase

Find the phase currents in the system.

Solution

(a) Since $V_P = \dfrac{V_L}{\sqrt{3}}$ for a star connected system having a neutral, then:

$$V_P = \frac{415}{\sqrt{3}} = \mathbf{240 \text{ V}}$$

and:

$$I_L = I_P = \frac{V_P}{R} = \frac{240}{50} = 4.8 \text{ A}$$

(b) Since $P = V_P I_P \cos \phi$, then:

$$I_P = \frac{P}{V_P \times \cos \phi}$$

$$= \frac{1\,000}{240 \times 0.8}$$

$$= 5.21 \text{ A}$$

(c) Since $Z = \sqrt{R^2 + X^2}$

$$= \sqrt{20^2 + 40^2}$$

$$= 44.72 \text{ }\Omega$$

then:

$$I_P = \frac{V_P}{Z} = \frac{240}{44.72} = 5.37 \text{ A}$$

158 A coil of copper wire has a resistance of 300 ohms at a temperature of 90 °C. Determine its resistance at a room temperature of 20 °C. Take the temperature coefficient of resistance of copper to be 0.004 3 Ω/Ω °C at 0 °C.

Solution

The formula to use in solving this problem is:

$$\frac{R_1}{R_2} = \frac{1 + at_1}{1 + at_2}$$

where $R_1 = 300 \text{ }\Omega$
$t_1 = 90$ °C
$t_2 = 20$ °C

By transposition:

$$R_2 = \left(\frac{1 + at_2}{1 + at_1}\right) R_1$$

$$= \left(\frac{1 + 0.086}{1 + 0.387}\right) \times 300$$

$$= 234.9 \text{ }\Omega$$

159 Draw a diagram showing how a single-phase wattmeter, voltmeter and ammeter are connected to a resistive load. Assume the use of a current transformer in view of the load current being 300 A.

Solution

See Figure 73.

160 (a) For the circuit shown in Figure 74 the value of R is 12 Ω and the value of X_L is 16 Ω. Calculate the:
   (i) impedance of the circuit
   (ii) current
   (iii) voltage across each component.
(b) Draw a phasor diagram showing the phase relation between the current, the supply voltage and the voltages across the resistor and inductor.

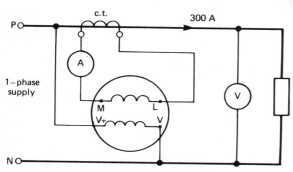

Figure 73 *Current transformer connection*

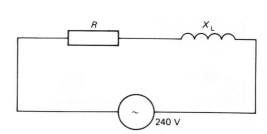

Figure 74 *Series RL circuit*

(c) From the phasor diagram, or otherwise determine:
  (i) the phase angle between current and supply voltage
  (ii) the power factor.

CGLI/II/1983

Solution

(a) (i) $Z = \sqrt{R^2 + X_L^2}$
$= \sqrt{12^2 + 16^2}$
$= 20\ \Omega$

(ii) $I = \dfrac{V}{Z} = \dfrac{240}{20} = 12\ \text{A}$

(iii) $V_R = IR = 12 \times 12 = 144\ \text{V}$
$V_L = IX_L = 12 \times 16 = 192\ \text{V}$

Check:
$V_S = \sqrt{V_R^2 + V_L^2}$
$= \sqrt{144^2 + 192^2}$
$= 240\ \text{V}$

(b) See Figure 75.

(c) (i) Use protractor or use formula:
$\cos \phi = \dfrac{R}{Z}$ and find phase angle (ϕ)
$= 53°$

(ii) Power factor $= \dfrac{R}{Z} = \dfrac{12}{20} = \mathbf{0.6\ lagging}$

161 From Figure 76, determine the value of the resistor marked X if the p.d. across the 2 Ω resistor is 60 V.

Solution

The current flowing through the circuit is found by:

$I = \dfrac{V_2}{R} = \dfrac{60}{2} = 30\ \text{A}$

Since $E = V_1 + V_2$

where E is the supply voltage
V_1 and V_2 are the circuit p.d.s
then $V_1 = E - V_2$
$= 240 - 60$
$= 180\ \text{V}$

The current through the 9 Ω resistor is:

$I = \dfrac{V_1}{R} = \dfrac{180}{9} = 20\ \text{A}$

The current through the unknown resistor is:

$30 - 20 = 10\ \text{A}$

The value of the unknown resistor is:

$R_X = \dfrac{V_1}{I} = \dfrac{180}{10} = 18\ \Omega$

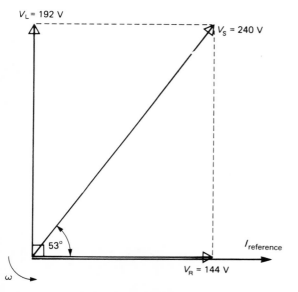

Figure 75 *Phasor diagram of RL series circuit*

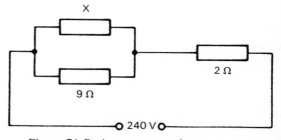

Figure 76 *Resistance connections*

162 A test run is required on a single-phase, 240 V motor of 4 kW rating.
(a) Draw a connection diagram showing instruments connected in the circuit to measure current, voltage and power.
(b) For each instrument used
  (i) name a type suitable for use in the circuit.
  (ii) state an acceptable range to enable satisfactory readings to be obtained assuming a full-load power factor of 0.8.
(c) State:
  (i) which of the instruments will require protection from the starting current of the motor.
  (ii) *one* method by which this protection may be achieved.
(d) During one test the following readings were taken:

	¼ full-load	full-load
voltage (volts)	240	240
current (amperes)	10	20
power (watts)	1000	4000

Calculate the power factor at each of these loads.

CGLI/II/88

Solution

(a) A diagram of this connection is shown in Figure 65.
(b) (i) Since the supply is a.c., moving iron instruments can be used for both the ammeter and voltmeter. A suitable wattmeter instrument for measuring the power would be a dynamometer.
  (ii) At full-load the motor will be taking 20.83 A and a suitable ammeter would be rated 0/30 A. The voltmeter would need to be rated at 0/250 V and the wattmeter, while only measuring 4 kW, would be suitably rated at 0/5 kW.

(c) (i) The in-rush current to the motor's windings is about 6–8 times full-load current and the instrument that requires protection is the ammeter.
  (ii) One method by which protection is achieved is to place a short-circuiting switch across the ammeter's terminals. After the motor has run up to speed the switch can be opened.
(d) The power factor at ¼ full-load is:
Power factor = watts/voltamperes
= 1000/2400
= **0.416 lagging**
The power factor at full-load is:
Power factor = watts/voltamperes
= 4000/4800
= **0.833 lagging**
Note: The power factor of the motor improves as more load is placed on it.

163 Draw a phasor diagram of Figure 77 and find the value of supply current and power factor.

Solution

See Figure 78.

I_S = **10 A** leading the voltage by 30°
Power factor (cos ϕ) = cos 30° = **0.866 leading**

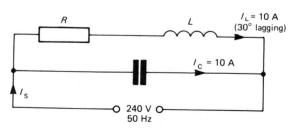

Figure 77 *Series/parallel RLC circuit*

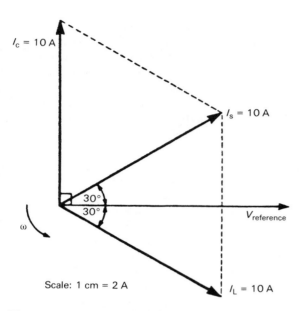

Figure 78 *Phasor diagram of Figure 77*

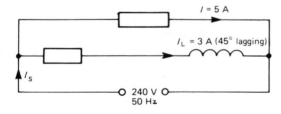

Figure 79 *Series/parallel RL circuit*

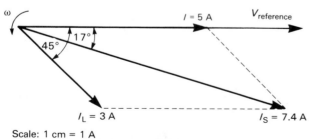

Scale: 1 cm = 1 A

Figure 80 *Phasor diagram of Figure 79*

164 Draw a phasor diagram of Figure 79 and determine the supply current and power factor.

Solution

See Figure 80.

$I_S = 7.4$ **A** (by measurement) lagging the voltage by 17°

Power factor $(\cos \phi) = \cos 17° = $ **0.956 lagging**

165 (a) Draw a circuit diagram showing how a voltage transformer and current transformer are connected in a single-phase supply system so that a wattmeter can be used to measure power. Assume the system to be at a high voltage and taking a large load current.
(b) Briefly state some advantages of using a current and voltage instrument transformer.

Solution

(a) See Figure 81.
(b) A current transformer is used where the circuit current is very high. Its secondary connection can be taken to a normal size instrument using much smaller cables, allowing the instrument to be read some distance from the actual measuring point.

A voltage transformer is used to measure a proportion of the system's high voltage, again using much smaller size cables. It is usual to find the secondary windings of current transformers catering for 5 A circuits and voltage transformer secondary windings catering for 110 V circuits.

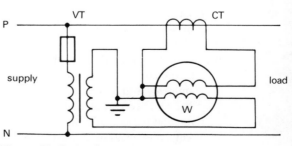

Figure 81 *C.t. and v.t. connections*

166 An auto-transformer for a single-phase operation has 600 turns. The primary supply of 240 V is connected to 450 turns while the secondary load is connected at 300 turns.
(a) Show a diagram of the arrangement.
(b) Determine the secondary voltage.
(c) Ignoring losses, if the secondary load current is 10 A what current is taken from the supply?
(d) What is the value of current in the common section of the transformer?

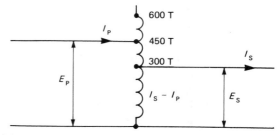

Figure 82 *Autotransformer connections*

Solution

(a) See Figure 82.

(b) $E_s = E_p \times \dfrac{N_s}{N_p}$

$= 240 \times \dfrac{300}{450}$

$= \mathbf{160\ V}$

(c) $I_p = \dfrac{N_s I_s}{N_p}$

$= \dfrac{300 \times 10}{450}$

$= \mathbf{6.66\ A}$

(d) Current in common section is:

$I_s - I_p = 10 - 6.66 = \mathbf{3.34\ A}$

167 Draw a circuit diagram showing how an ammeter and selector switch are connected via current transformers in a 3-phase, 3-wire system in order for the ammeter to measure current in any line.

Solution

See Figure 83.

Note: It should be pointed out that it is possible for a dangerously high voltage to be induced in the secondary winding of a current transformer if its circuit is left open. In view of this, the selector switch must be capable of shorting out those current transformers not in use.

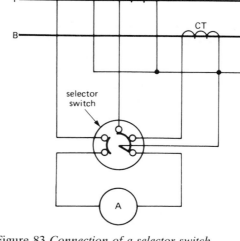

Figure 83 *Connection of a selector switch*

168 A moving coil milli-ammeter has a resistance of 5 Ω with full scale deflection at 15 mA. The scale reads 0–15 mA with 15 divisions.
(a) With the aid of circuit diagrams explain how:
  (i) the current range of the instrument may be extended
  (ii) the milli-ammeter can be adapted for use as a voltmeter.
(b) (i) Calculate the value of the resistor required to enable the milli-ammeter to read 0–3 A.

(ii) By what factor must the scale reading be multiplied to give correct readings of current?

(c) (i) Calculate the value of the resistor required to enable the milli-ammeter to read 0–150 V.

(ii) How many volts will one division of the scale now represent?

CGLI/II/80

Solution

(a) See Figure 84.
(i) To extend the range of the milli-ammeter to read more current, a resistor called a *shunt* is used. This component is wired or fitted across the milli-ammeter's terminals (i.e. *shunted* across). The shunt resistor is very often of a low ohmic value since it allows a large proportion of the circuit current to pass. The instrument will not be damaged provided the correct value of shunt is determined. The scale reading will also need modifying.

(ii) To extend the range of the milli-ammeter to read voltage, a resistor called a *multiplier* is used. This component is wired or fitted in series with the instrument. The multiplier resistor is very often of a high ohmic value in order to limit the current flowing through the instrument up to full scale deflection. The scale will have to be changed so that it reads voltage.

(b) The calculations are as follows:
(i) If the instrument has to be modified to read 3 A, then the shunt resistor must take the difference between this reading and the f.s.d. value of 15 mA, i.e. $3 - 0.015 = 2.985$ A.

Since the potential difference will be the same across the shunt as it is across the instrument for f.s.d. then finding the p.d. will help find the value of the shunt. Thus:

$$\text{p.d.} = I_{\text{f.s.d.}} \times R_{\text{inst.}}$$
$$= 0.015 \times 5$$
$$= 75 \text{ mV}$$

$$\text{Shunt resistor} = \frac{\text{p.d.}}{\text{shunt current}}$$
$$= \frac{0.075}{2.985}$$
$$= \mathbf{0.025 \ \Omega}$$

(ii) The factor required is found by simply dividing the instrument's f.s.d. reading into 3 A, i.e.

$$\frac{3}{0.015} = \mathbf{200}.$$

(c) (i) The value of multiplier resistor is found by firstly finding the p.d. across it, then dividing by the instrument's f.s.d. of 15 mA. Thus:

$$\text{p.d.} = 150 - 0.075$$
$$= 149.925 \text{ V}$$

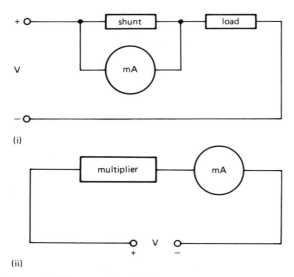

Figure 84 *Shunt and multiplier connections*

Multiplier resistor $= \dfrac{\text{p.d.}}{I_{\text{f.s.d.}}}$

$= \dfrac{149.925}{0.015}$

$= 9\ 995\ \Omega$

(ii) Each scale division of the instrument will represent

$\dfrac{150\ \text{V}}{15} = \textbf{10 volts}$

169 (a) State the conditions necessary to produce resonance in the circuit shown in Figure 85.

(b) If the resistance of $R = 8\ \Omega$ and the inductive reactance of $L = 15\ \Omega$ determine the value of capacitance for resonance to occur.

Solution

(a) For resonance to occur X_L must equal X_C.

(b) Since $X_L = 15\ \Omega$, then $X_C = 15\ \Omega$

Thus $X_C = \dfrac{1}{2\pi fC}$

By transposition:

$C = \dfrac{10^6}{2\pi F X_C}\ \mu\text{F}$

$= \dfrac{10^6}{314.2 \times 15}$

$= \textbf{212}\ \boldsymbol{\mu}\textbf{F}$

170 If the capacitor in Figure 85 had a value of 100 μF and the supply voltage was 240 V determine:
(a) the circuit current if R and X_L were the same as before in question 169
(b) the power factor of the circuit.

Solution

(a) Since $X_C = \dfrac{10^6}{2\pi fC}$

$= \dfrac{10^6}{314.2 \times 100}$

$= 31.8\ \Omega$

Since $Z = \sqrt{R^2 + (X_C - X_L)^2}$

$= \sqrt{8^2 + (31.8 - 15)^2}$

$= \sqrt{64 + 282.24}$

$= 18.6\ \Omega$

Since $I = \dfrac{V}{Z}$

$= \dfrac{240}{18.6}$

$= \textbf{12.9 A}$

(b) Power factor $= \dfrac{R}{Z}$

$= \dfrac{8}{18.6}$

$= \textbf{0.43 leading}$

171 A 6.8 kW/415 V/50 Hz, three-phase induction motor has an efficiency of 85% and a power factor of 0.7 lagging. Determine the value of power factor correction capacitors to improve the power factor to unity condition.

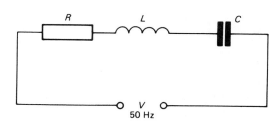

Figure 85 *Series RLC circuit*

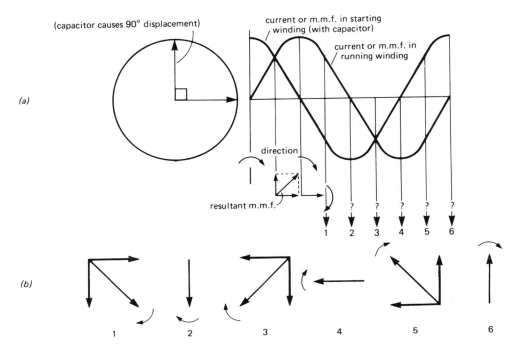

Figure 86 *Split-phase rotation*

Solution

Since efficiency (p.u.) = $\dfrac{\text{output}}{\text{input}}$

$$\text{input} = \dfrac{\text{output}}{\text{efficiency (p.u.)}}$$

$$= \dfrac{6.8}{0.85} = 8 \text{ kW}$$

Since kilovoltamperes = $\dfrac{\text{kW}}{\text{p.f.}}$

$$= \dfrac{8}{0.7} = 11.43 \text{ kVA}$$

The lagging reactive $\text{kVA}_r = \sqrt{\text{kVA}^2 - \text{kW}^2}$
$= \sqrt{11.43^2 - 8^2}$
$= \mathbf{8.16 \text{ kVA}}$

For unity power factor conditions, the injected leading capacitive kVA_r must equal the lagging reactive kVA_r of 8.16 kVA. Thus:

$$I_L = \dfrac{\text{kVA}_r}{\sqrt{3}V_L}$$

$$= \dfrac{8160}{\sqrt{3.415}}$$

$$= 11.35 \text{ A}$$

Since $I_P = \dfrac{I_L}{\sqrt{3}}$

$$= \dfrac{11.35}{\sqrt{3}}$$

$$= 6.55 \text{ A}$$

Then $X_C/\text{phase} = \dfrac{V_L}{I_P}$

$$= \dfrac{415}{6.55}$$

$$= 63.35 \ \Omega$$

Thus $C/\text{phase} = \dfrac{10^6}{314.2 \times 63.35}$

$$= \mathbf{50 \ \mu F}$$

Science and Calculations

172 Using Figure 86 (a), complete the resultant m.m.f. rotating magnetic field direction.

Solution

See Figure 86(b).

173 Draw a diagram of a single-phase a.c. motor controlled by a direct on-line contactor starter having thermal overloads. The circuit should include remote start and stop buttons.

Solution

See Figure 87.

174 Draw block and circuit diagrams of controlgear for a 3-phase induction motor. State on the diagrams the size of cable and fuse protection assuming the motor will take a current of 15 A.

Solution

See Figure 88.

175 What are the IEE Wiring Regulations covering motor circuits?

Solution

Reference should be made to Chapter 55, Section 552. Equipment and cables feeding motors should be rated to carry the full load current of the motors. The supply undertaking should be consulted regarding starting arrangements for those motors requiring heavy starting currents. Motors exceeding 0.37 kW require control equipment incorporating means of protection against overcurrent, although if a motor is part of some current-using equipment complying with a British Standard as a whole, it is exempt from such protection.

Every electric motor must be provided with a means to prevent automatic restarting which could cause a danger. The provision of *no-volt protection* in the starter/control equipment may meet this requirement. Where a motor is started and stopped at intervals by an automatic control device, the requirements do not apply, providing other adequate measures

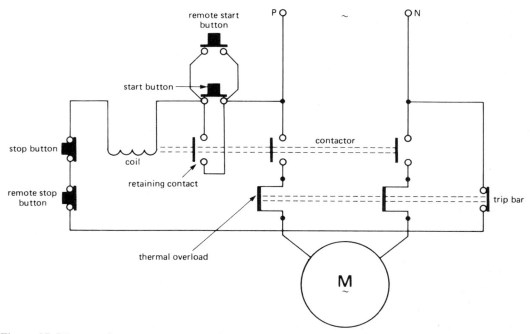

Figure 87 *Direct-on-line contactor starter for single-phase motor*

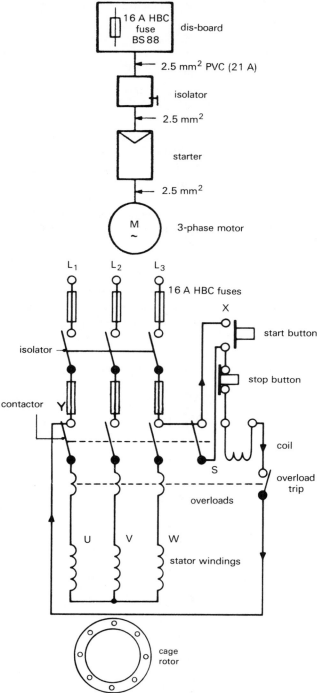

Figure 88 *3-phase cage induction motor starting arrangements (a) Direct-on-line contactor (b) Motor controlgear*

are taken against danger of unexpected restarting.

176 Determine the synchronous speed (n_s) and rotor speed (n_r) from the following induction motor data:
(a) $f = 50$, $p = 2$ and $s = 0.03$
(b) $f = 50$, $p = 4$ and $s = 0.03$
(c) $f = 50$, $p = 6$ and $s = 0.03$
(d) $f = 50$, $p = 6$ and $s = 0.05$
(e) $f = 60$, $p = 6$ and $s = 0.05$

Note: f is the frequency of supply in hertz
p is the number of stator poles
s is the slip in per unit values

Solution

(a) $n_s = \dfrac{f}{p}$ $\qquad n_r = n_s(1 - s)$

$ = \dfrac{50}{1}$ $\qquad = 50(1 - 0.03)$

$ = \mathbf{50\ rev/s}$ $\qquad = \mathbf{48.5\ rev/s}$

(b) $n_s = \dfrac{50}{2}$ $\qquad n_r = 25(1 - 0.03)$

$ = \mathbf{25\ rev/s}$ $\qquad = \mathbf{24.25\ rev/s}$

(c) $n_s = \dfrac{50}{3}$ $\qquad n_r = 16.66(1 - 0.03)$

$ = \mathbf{16.66\ rev/s}$ $\qquad = \mathbf{16.17\ rev/s}$

(d) $n_s = \dfrac{50}{3}$ $\qquad n_r = 16.66(1 - 0.05)$

$ = \mathbf{16.6\ rev/s}$ $\qquad = \mathbf{15.83\ rev/s}$

(e) $n_s = \dfrac{60}{3}$ $\qquad n_r = 20(1 - 0.05)$

$ = \mathbf{20\ rev/s}$ $\qquad = \mathbf{19\ rev/s}$

177 (a) What is an interpole? Explain where it is connected in a machine.
(b) Draw a diagram of a d.c. machine's main salient pole showing both series and shunt field connections to give cumulative compound characteristics. Show current directions in the field windings and pole polarity.

Solution

(a) An interpole is a small pole fitted between the main poles of a d.c. machine in order to neutralize the effects of armature reaction which causes sparking at the brushes. Interpoles are connected in series with the armature connections.

(b) See Figure 89, which is a main pole, not an interpole.

178 (a) Explain briefly how the back e.m.f. and the current change during the starting of a d.c. motor.

(b) A 200 V shunt-wound motor has an armature resistance of 0.25 Ω and a field resistance of 200 Ω. The motor gives an output of 4 kW at an efficiency of 80%. For this load calculate:
  (i) the motor power input in kW
  (ii) the load current
  (iii) the motor field current
  (iv) the armature current
  (v) the back e.m.f.

(c) What are the IEE requirements for the rating of fuses protecting a circuit feeding a motor?

CGLI/II/81

Solution

(a) The back e.m.f. of a d.c. motor is a generated e.m.f. created by the armature conductors cutting the main field as the armature revolves. This induced e.m.f. acts in opposition to the supply voltage and is always less than the supply voltage – the difference being the armature voltage drop.

As the armature accelerates, its back e.m.f. increases. Also, as the load current increases, the back e.m.f. decreases. If the armature stops revolving there will be no back e.m.f. and with the supply switched on, excessive current will be drawn by the motor.

(b) (i) Since $\text{efficiency} = \dfrac{\text{output}}{\text{input}}$

$$\text{input} = \dfrac{\text{output}}{\text{efficiency (p.u.)}}$$

$$= \dfrac{4\,000}{0.8}$$

$$= 5\,000 \text{ W}$$

$$= \mathbf{5 \text{ kW}}$$

(ii) $I_L = \dfrac{\text{power input}}{\text{supply voltage}}$

$$= \dfrac{5\,000}{200}$$

$$= \mathbf{25 \text{ A}}$$

(iii) $I_f = \dfrac{\text{supply voltage}}{\text{field resistance}}$

$$= \dfrac{200}{200}$$

$$= \mathbf{1 \text{ A}}$$

(iv) Since $I_L = I_a + I_f$
then $I_a = I_L - I_f$
$$= 25 - 1$$
$$= \mathbf{24 \text{ A}}$$

(v) $E_b = V - I_a R_a$
$$= 200 - (24 \times 0.25)$$
$$= \mathbf{194 \text{ V}}$$

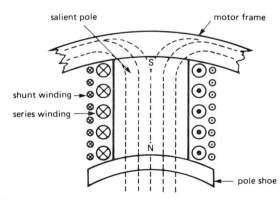

Figure 89 *Salient pole electromagnet*

(c) Reference should be made to the IEE

Wiring Regulations (15th Edition), Section 434, Reg. 434–5. In the 14th Edition, Reg. A–68 allowed fuses to be rated up to twice that of the cables between the fuse and the starter provided the starter afforded overload protection. However, the new requirements allow the use of an overload device complying with Section 433 to protect the conductors on the load side of the device, provided that it has a rated breaking capacity not less than the prospective short circuit current at the point of installation.

Note: While it is not required in the question, Figure 90 is a diagram of a shunt motor showing some of the terms mentioned in the solution above.

179 (a) Describe briefly the following parts of a d.c. motor:
  (i) armature
  (ii) commutator
  (iii) field system
(b) Draw a circuit diagram for:
  (i) a d.c. series motor
  (ii) a d.c. compound motor

CGLI/11/82

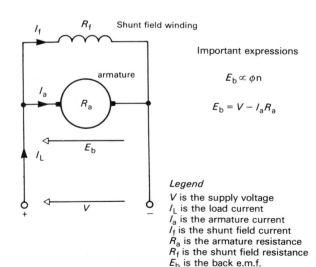

Figure 90 *D.c. shunt motor connections*

Solution

(a) (i) The armature is the name for the rotating part of the machine and is made up of many laminations of soft-magnetic-alloy material into which armature coils are assembled.
  (ii) The commutator is part of the armature, serving the purpose of transferring an external current to the armature conductors via brush-gear.
  (iii) The field system is that part of the d.c. motor which produces the excitation flux. It comprises the main poles and field windings which identify the machine as a *series*, *shunt* or *compound motor*.
(b) (i) See Figure 91(a).
  (ii) See Figure 91(b).

180 The speed of a 220 V d.c. motor with an armature current of 10 A and armature resistance 0.5 Ω is 12.5 rev/s. What would be its speed if the armature current was increased to 30 A? Assume the armature voltage (applied voltage) and field current remain unchanged.

Solution

Since $E = V - I_a R_a$

where E is the back e.m.f.
  V is the applied volts
  I_a is the armature current
  R_a is the armature resistance

First condition:
$$E_1 = 220 - (10 \times 0.5) = 215 \text{ V}$$

Second condition:
$$E_2 = 220 - (30 \times 0.5) = 205 \text{ V}$$

The back e.m.f. is proportional to the speed of the armature and magnetic flux per pole. If the field current remains unchanged, the conditions can be represented by the expression:

$$E \propto \phi n$$

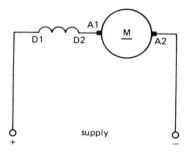

(a)

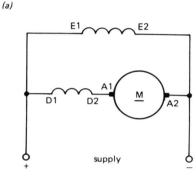

(b)

A1 – A2 armature connections
D1 – D2 series field connections
E1 – E2 shunt field connections

Figure 91 *D.c. motor connections*

(See page 80 of *Electrical Installation Technology 2*.)

Thus $E_1 \propto n_1$
$E_2 \propto n_2$

By proportion $n_2 = \dfrac{n_1 \times E_2}{E_1}$

$= \dfrac{12.5 \times 205}{215}$

$= \mathbf{11.9 \ rev/s}$

181 Determine the efficiency and power factor of a single-phase motor having the following data:
Electrical input – wattmeter reading 13 120 W
voltmeter reading 240 V
ammeter reading 74 A
Mechanical output – 11 190 W

Solution

$$\text{efficiency} = \dfrac{\text{output}}{\text{input}}$$

$$= \dfrac{11\ 190}{13\ 120}$$

$$= \mathbf{0.85 \ p.u. \ (85\%)}$$

$$\text{power factor} = \dfrac{\text{input power}}{\text{voltamperes}}$$

$$= \dfrac{13\ 120}{240 \times 74}$$

$$= \mathbf{0.74 \ lagging}$$

182 Determine the power output, power factor and efficiency of a three-phase motor having the following test data:
1 speed 23.75 rev/s
2 input wattmeter reading 16 920 W
3 voltmeter reading 400 V
4 ammeter reading 39.5 A
5 brake pulley diameter 0.33 m
6 effective pull at circumference of pulley 564.44N

Solution

Power output (mechanical) is given by the expression:

$P_o = 2\pi.nT$ watts
$= 2 \times 3.142 \times 23.75 \times (564.44 \times 0.165)$
$= 13\ 899$ W or **13.9 kW**

Note: Torque (T) = Force (N) × radius of pulley (m)

$$\text{efficiency} = \dfrac{\text{output}}{\text{input}}$$

$$= \dfrac{13\ 899}{16\ 920}$$

$$= \mathbf{0.82 \ p.u. \ (82\%)}$$

and input (electrical) is given by the expression:

$$P_i = \sqrt{3} V_L I_L \cos \phi \text{ watts}$$

where $\cos \phi$ is the power factor
By transposition:

$$\cos\phi = \frac{P_i}{\sqrt{3}V_L I_L}$$

$$= \frac{16\,920}{1.732 \times 400 \times 39.5}$$

$$= \mathbf{0.62\ lagging}$$

183 The current taken by a 240 V, 50 Hz single-phase induction motor is 39 A at a power factor lagging of 0.75. Determine:
(a) the input power in kilowatts (P)
(b) the kilovolt-amperes (S)
(c) the size of capacitor which will raise the power factor to unity

Solution

(a) $P = VI\cos\phi$

$$= 240 \times 39 \times 0.75$$

$$= 7\,020\ \text{W}$$

$$= \mathbf{7.02\ kW}$$

(b) $S = 240 \times 39$

$$= 9\,360\ \text{VA}$$

$$= \mathbf{9.36\ kVA}$$

Hence $Q = \sqrt{S^2 - P^2}$

$$= \sqrt{9.36^2 - 7.02^2}$$

$$= \sqrt{38.33}$$

$$= 6.19\ \text{kVA}$$

therefore $I_c = \dfrac{VI_r}{V}$

$$= \frac{6\,190}{240}$$

$$= 25.79\ \text{A}$$

and $X_c = \dfrac{V}{I_c}$

$$= \frac{240}{25.79}$$

$$= 9.3\ \Omega$$

thus $C = \dfrac{10^6}{2 \times 50 \times 9.3}$

$$= \mathbf{342\ \mu F}$$

(c) $C = \dfrac{10^6}{2\pi f X_c}\ \mu\text{F}$ but $X_c = \dfrac{V}{I_c}$

where X_c is the capacitive reactance
V is the supply voltage
I_c is the capacitive current

Since V is known, I_c must be found. The usual way to do this is by the formula:

$$I_c = \frac{VI}{V}$$

Here VI represents the reactive voltamperes (or leading VA injected by the capacitors) to bring power factor to unity. For simplicity see Figure 92.

184 (a) Explain with the aid of diagrams the method of reversing the direction of rotation of *each* of the following types of motor:
(i) three-phase induction
(ii) single-phase capacitor start
(iii) d.c. shunt
(iv) series

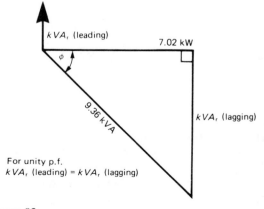

Figure 92

Science and Calculations

(b) Calculate the full load current of a 48 kW 3-phase 415 V motor given that the efficiency and power factor at full load is 85% and 0.9 respectively.

CGLI//II/83

Solution

(a) Methods of reversing the direction of rotation of motors are given on page 84 of *Electrical Installation Technology 2*.
  (i) Change any two supply leads.
  (ii) Change connections of running winding *or* starting winding.
  (iii) Change connections of shunt field winding *or* armature winding.
  (iv) Change connections of series field winding *or* armature winding.
  See Figure 93.

(b) $\text{efficiency} = \dfrac{\text{output}}{\text{input}}$

By transposition:

$$\text{input} = \dfrac{\text{output}}{\text{efficiency}}$$

$$= \dfrac{48\,000}{0.85}$$

$$= 56.47 \text{ kW}$$

Since input $(P) = \sqrt{3} V_L I_L \cos \phi$

By transposition:

$$I_L = \dfrac{P}{\sqrt{3} V_L \cos \phi}$$

$$= \dfrac{56\,470}{\sqrt{3} \times 415 \times 0.9}$$

$$= 87.29 \text{ A}$$

185 A 1 kW, 240 V, 50 Hz, 2 pole single-phase induction motor operates with 5% slip, 75% efficiency and 0.7 power factor on full load.
  (a) Draw a labelled circuit diagram of a push button starter with undervoltage and overcurrent protection for the above motor.

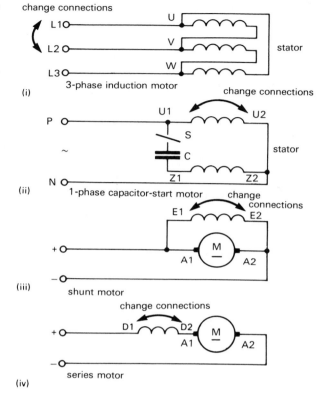

Figure 93 *Motor reversal of direction*

  (b) For full load conditions calculate the:
    (i) input power and current
    (ii) motor speed

CGLI/II/83

Solution

(a) See Figure 94.

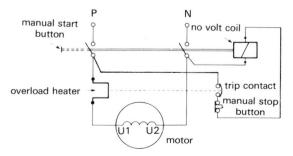

Figure 94 *Single-phase motor connections*

(b) (i) Since

$$\text{efficiency} = \frac{\text{output}}{\text{input}}$$

then

$$\text{input} = \frac{\text{output}}{\text{efficiency (p.u.)}}$$

$$= \frac{1\,000}{0.75}$$

$$= 1.33 \text{ kW}$$

because

$$\text{input} = V_p I_p \cos \phi$$

By transposition:

$$I = \frac{P}{V_p \cos \phi}$$

$$= \frac{1\,333.3}{240 \times 0.7}$$

$$= \mathbf{7.94 \text{ A}}$$

(ii) Since $n_s = \dfrac{f}{p}$

$$= \frac{50}{1}$$

$$= 50 \text{ rev/s}$$

where n_s is the synchronous speed
f is the supply frequency
p is the number of pairs of poles

and $n_r = n_s(1 - s)$

where n_r is the rotor speed
s is the per unit slip

Thus: $n_r = 50(1 - 0.05)$

$$= \mathbf{47.5 \text{ rev/s}}$$

186 A 50 kW a.c. motor operates at a power factor of 0.65 lagging and an efficiency of 83%. A 40 kVA$_r$ power factor improvement capacitor is connected in parallel with the motor. Find graphically or by calculation the:
(a) full load input kVA to the motor

(b) supply kVA and power factor when the motor is run on
(i) full load
(ii) no load   CGLI/II/82

Solution

(a) Motor's input $= \dfrac{\text{output}}{\text{efficiency (p.u.)}}$

$$= \frac{50}{0.83}$$

$$= 60.24 \text{ kW}$$

motor's VI $= \dfrac{P}{\text{p.f.}}$

$$= \frac{60.24}{0.65}$$

$$= 92.677 \text{ kVA}$$

The motor's lagging reactive component is:

$$Q = \sqrt{S^2 - P^2}$$

$$= \sqrt{92.677^2 - 60.24^2}$$

$$= 70.43 \text{ kVA}_r$$

Since 40 kVA$_r$ is injected, the remaining lagging kVA$_r$ is **30.43 kVA**

(b) (i) Full load $S = \sqrt{P^2 + Q^2}$

$$= \sqrt{60.24^2 + 30.43^2}$$

$$= \mathbf{67.49 \text{ kVA}}$$

(ii) No load conditions assumes that no capacitors are connected and in this case the kilovoltamperes = **92.677 kVA**

Note: Figure 95 is a graphical solution. Also, if the motor is a cage induction motor, it will have a poor power factor on no-load.

187 A 6-pole, 415 V, 50 Hz, three-phase induction motor operates at 0.7 power factor lagging and drives an elevator lifting 100 kg at a rate of 3.6 m/sec. If the elevator has an efficiency of 75% and the motor an efficiency of 85%, determine for full load conditions:

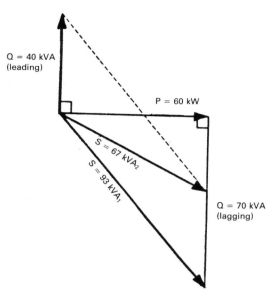

Figure 95 *Power factor correction for an a.c. motor*

(a) the motor's output and input
(b) the motor's line current and phase current, assuming the windings are delta connected
(c) the motor's synchronous speed and rotor speed assuming a 0.05 p.u. slip

Solution

(a) Work done/second = force × distance
= 100 × 3.6
= 360 kgf

If 1 kgf = 9.81 N, then power required by elevator is:

9.81 × 360 = 3 531.6 W

motor's output = elevator's input (see Figure 96)

$$P_o = \frac{3\,531.6 \times 100}{75}$$

= **4 708.8 W**

motor's input = $\frac{4\,708.8 \times 100}{85}$

(Pi) = **5 539.76 W**

(b) Since $P_i = \sqrt{3} V_L I_L \cos \phi$

then $I_L = \dfrac{P_i}{\sqrt{3} V_L \cos \phi}$

$= \dfrac{5\,539.76}{\sqrt{3} \times 415 \times 0.7}$

= **10.13 A**

$I_P = \dfrac{I_L}{\sqrt{3}}$

$= \dfrac{10.13}{\sqrt{3}}$

= **5.85 A**

(c) $n_s = \dfrac{f}{p}$

$= \dfrac{50}{3}$

= **16.66 rev/s**

$n_r = n_s(1 - s)$

= 16.66(1 − 0.05)

= **15.83 rev/s**

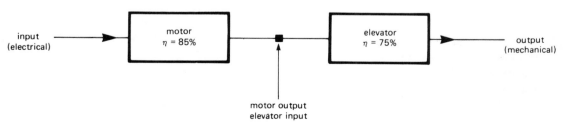

Figure 96 *Machine efficiencies*

188 A machine is driven at 10 rev/s by a belt from a motor which is running at a speed of 25 rev/s. If the motor is fitted with a pulley of diameter 176 mm, find the size of pulley for the machine and belt speed assuming no slip.

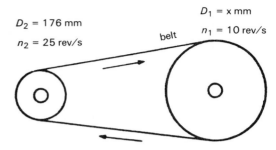

Figure 97 *Belt connections*

Solution

The arrangement is shown in Figure 97. The expression used for the relationship between motor and machine pulley speeds and diameters is:

$$\frac{\text{speed of driven pulley }(n_1)}{\text{speed of driver pulley }(n_2)} = \frac{\text{diameter of driver pulley }(D_2)}{\text{diameter of driven pulley }(D_1)}$$

In this case:
$$n_1 = 10 \text{ rev/s}$$
$$n_2 = 25 \text{ rev/s}$$
$$D_2 = 176 \text{ mm}$$

Thus:
$$D_1 = \frac{D_2 \times n_2}{n_1}$$
$$= \frac{176 \times 25}{10}$$
$$= 440 \text{ mm}$$

speed of belt = π × diameter × speed of pulley
= π × 176 × 25 (using motor data)
= 13 824.8 mm/s
= **13.8 m/s**

189 Explain what happens when forward and reverse bias is applied to a pn junction diode.

Solution

Reference should be made to the 'Diode', page 96 in *Electrical Installation Technology 2*. Figure 98 shows two conditions created by the battery connections. The *forward bias* connection is when the positive battery terminal is connected to the 'p' material while the negative battery terminal is connected to the 'n' material. If the battery voltage is high enough it will destroy the internal barrier potential and allow electrons to pass into the 'p' material and flow around the circuit (electron flow). The pn junction diode is now in its 'on' state.

If the polarity of the battery connections to the diode is reversed as shown, then the electrons in the 'n' material are attracted towards the positive battery connection and holes in the 'p' material are attracted towards the negative battery connection. This creates a wide depletion layer and very little current flows around the circuit. The diode is said to have *reverse bias* and is now in its 'off' state.

Note: When connected to a d.c. supply, the PN diode acts as a switch, either 'on' or 'off' depending on its connection with the supply terminals.

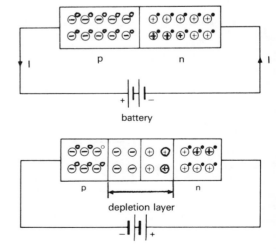

Figure 98 *Forward and reverse bias*

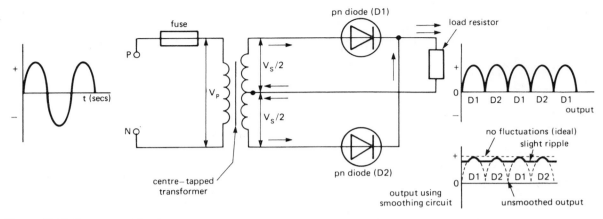

Figure 99 *Full-wave rectification*

190 Explain how full-wave rectification is obtained from a single-phase a.c. supply. Draw the a.c. voltage waveform and d.c. output voltage waveform.

Solution

There are commonly two methods of achieving full-wave rectification using pn junction diodes, (a) using a centre-tapped transformer and (b) using a bridge rectifier. Figure 99 shows the use of two diodes connected to a centre-tapped transformer.

When diode D1 is positive, it will conduct in the forward direction through the load and back via the centre point of the transformer. The signal down to diode D2 is blocked. In the second half cycle, D2 conducts while D1 blocks and the output signal takes on the pattern shown. An improvement to the output signal is by using a smoothing circuit which raises the level of signal, making it steady with very little ripple.

191 What is a heat sink?

Solution

Solid state devices, such as diodes and thyristors, while being robust in design may become damaged by excessive heat and temperature rise beyond their working performance. Because of this they are fitted with heat sinks as part of their assembly. These are usually a mass of metal, much larger than the device need be, with the purpose of dissipating any heat away from the device. Heat sinks may take various shapes and sizes, and devices fitted with cooling fins to increase the area of their radiating surface are quite common. See Figure 100.

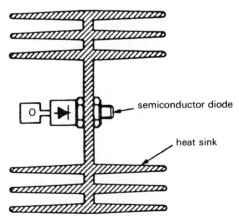

Figure 100 *Heat sink*

(i)

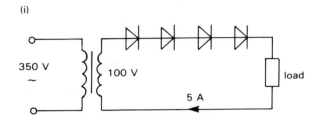

(ii)

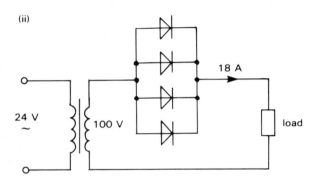

(iii)

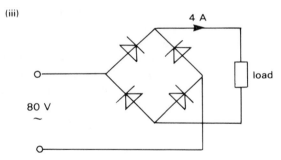

Figure 101 *Diode connections*
(i) Half-wave rectification using step-down transformer
(ii) Half-wave rectification using step-up transformer
(iii) Full-wave rectification

192 The following question relates to semiconductor diodes.
  (a) Draw a circuit diagram to show how four semiconductor diodes, each rated at 100 V, 5 A, may be used to feed the feed the following d.c. loads:
   (i) half-wave d.c. 5 A output from a 350 V a.c. input
   (ii) half-wave d.c. 18 A output from a 24 V a.c. input
   (iii) full-wave d.c. 4 A output from an 80 V a.c. input.
  (b) Describe how a diode may be tested with the aid of an ohmmeter.
  (c) State a method used to dissipate the heat produced in a diode when in use.

CGLI/II/88

Solution

(a) See Figure 101.
(b) A diode is a rectifying device made to allow current to flow in one direction only, i.e. forward bias. It will block current flow in the other direction, i.e. reverse bias. The ohmmeter, when connected to the diode, is switched to its continuity scale and will read a very low ohmic value when the diode is connected in the forward mode. Changing over the connections will show a high reading since it is now in the reverse mode.
(c) The diode is a solid state device and requires the assistance of a heat sink to dissipate the circuit's heating effect.

Note: A diagram of this device is shown in Figure 100.

193 Determine the illuminance on the surface directly below an incandescent lamp of 1 200 cd if the distance is:
(a) 3 m
(b) 6 m
(c) 9 m

Solution

(a) $E = \dfrac{I}{d^2}$ lux

$= \dfrac{1\,200}{9}$

$= \mathbf{133.3\ lx}$

(b) $E = \dfrac{1\,200}{36}$

$= \mathbf{33\ lx}$

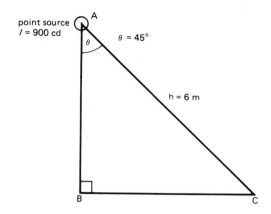

Figure 102

(c) $E = \dfrac{1\,200}{81}$

$= 14.8 \text{ lx}$

194 With reference to Figure 102, determine the illuminance at point C.

Solution

See Figure 102.

$$E = \dfrac{I}{h^2}\cos \phi$$

where h is the distance 6 m

$$E = \dfrac{900}{36} \times 0.707$$

$= 17.67 \text{ lx}$

195 Calculate the average illuminance in a room measuring 10.5 m by 7 m, assuming the light output of each luminaire lighting the room to be 5 000 lm with utilization factor and maintenance factor 0.5 and 0.8 respectively. Sixteen luminaires are installed.

Solution

The average illuminance required is given by the expression:

$$E = \dfrac{F \times MF \times CU}{A}$$

where F is the lumen output of luminaires
MF is the maintenance factor
CU is the coefficient of utilization
A is the area required to be lit

$$E = \dfrac{5\,000 \times 16 \times 0.8 \times 0.5}{10.5 \times 7}$$

$= \textbf{435.4 lx}$

196 Two incandescent filament lamps of luminous intensity 100 cd and 60 cd respectively in all directions are fixed to the ends of a photometer bench as shown in Figure 103. A movable, double-sided, white matt screen is placed between the lamps with its opposite faces normal to the rays from the lamps. The face opposite the 60 cd lamp receives an illuminance of 26 lx and the other side receives an illuminance of 67 lx. Find:
(a) the distance each lamp is from the screen
(b) the illuminance on each side of the screen when it is placed halfway between the lamps

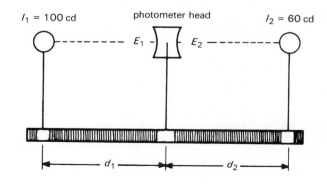

Figure 103

Solution

(a) Since $E = \dfrac{I}{d^2}$

By transposition: $d^2 = \dfrac{I}{E}$

Thus:
$$d_1 = \sqrt{\dfrac{I_1}{E_1}} = \sqrt{\dfrac{100}{67}} = 1.22 \text{ m}$$

and:
$$d_2 = \sqrt{\dfrac{I_2}{E_2}} = \sqrt{\dfrac{60}{26}} = 1.52 \text{ m}$$

(b) Since $= d_1 + d_2$
$= 1.22 + 1.52$
$= 2.74 \text{ m}$

The screen positioned halfway would be

$\dfrac{2.74}{2} = 1.37 \text{ m}$

Thus:
$$E_1 = \dfrac{I_1}{d_2} = \dfrac{100}{1.876\,9} = 53 \text{ lx}$$

and:
$$E_2 = \dfrac{I_2}{d_2} = \dfrac{60}{1.876\,9} = 32 \text{ lx}$$

197 With reference to Figure 103, assuming the distance between the lamps to be 2.74 m, what position has the screen to be moved to in order to provide equal illumination on both sides?

Solution

In this case $\quad E_1 = E_2$

i.e. $\quad \dfrac{I_1}{d_1^2} = \dfrac{I_2}{d_2^2}$

By transposition: $\quad \dfrac{I_1}{I_2} = \left(\dfrac{d_1}{d_2}\right)^2$

Thus: $\quad \sqrt{\dfrac{100}{60}} = \dfrac{d_1}{d_2}$

$1.29 = \dfrac{d_1}{d_2}$

$1.29 d_2 = d_1$

Since $\quad d = d_1 + d_2$
$= 2.74 \text{ m}$
$d_1 = 2.74 - d_2$

There are now two conditions for d_1

$1.29 d_2 = 2.74 - d_2$
$d_2 + 1.29 d_2 = 2.74$
$2.29 d_2 = 2.74$
$d_2 = \dfrac{2.74}{2.29}$
$= \mathbf{1.196 \text{ m}}$

Therefore:
$d_1 = 2.74 - 1.196$
$= \mathbf{1.544 \text{ m}}$

198 Figure 104 shows the connections of a SON discharge lamp. With switch S open, ammeter A_1 reads 5 A and W reads 420 W. With S closed A_2 reads 2.2 A and A_1 reads 3 A, W reads 420 W again.
(a) Draw a phasor diagram of the circuit using 1 A = 2 cm.
(b) If the lamp emits 44 000 lm and losses in the circuit are 20 W, what is the efficacy of the lamp?

Solution

(a) See Figure 105.

Note: Since power factor $= \dfrac{P}{VI} = \dfrac{420}{240 \times 5}$
$= 0.35 \text{ lagging,}$

the phase angle (θ) is 69°30′ (approx.).

(b) The lamp is a 400 W SON and its efficacy is given by the expression:

$$\dfrac{\text{lumens}}{\text{watts}} = \dfrac{44\,000}{400} = \mathbf{110 \text{ lm/w}}$$

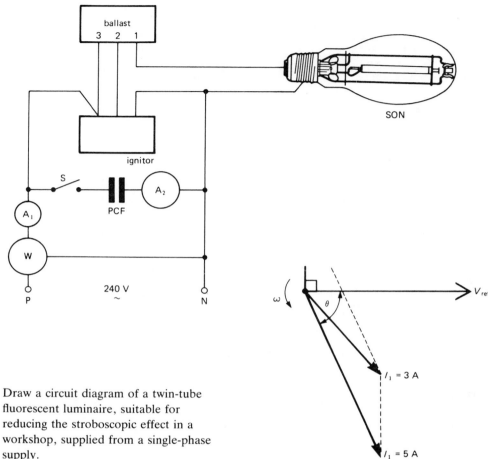

Figure 104

Figure 105

199 (a) Draw a circuit diagram of a twin-tube fluorescent luminaire, suitable for reducing the stroboscopic effect in a workshop, supplied from a single-phase supply.

(b) For the circuit shown in Fig. 2 (Figure 106) determine the current measured by:
  (i) meter A1
  (ii) meter A2
  (iii) meter A3

CGLI/II/88

Solution

(a) See Figure 107.

(b) (i) The impedance of RX in series is 50 ohm. The current through meter A1 is

$$I = V/Z = 240/50 = 4.8 \text{ A}$$

This current lags behind the supply voltage by a phase angle. This is determined as follows:

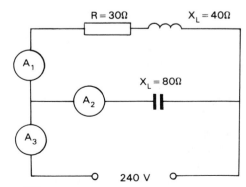

Figure 106

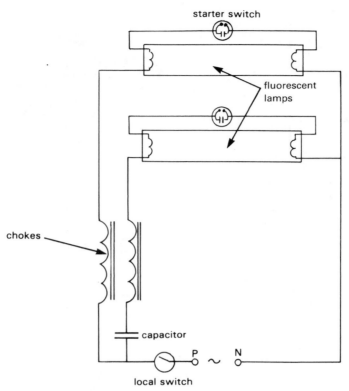

Figure 107 *Typical lead-lag circuit*

Since power factor (cos ϕ) is ratio R/Z, then cos ϕ = 30/50 = 0.6 lagging thus ϕ = 53° (approx.)

(ii) The current through meter A2 is

$I = V/X = 240/80 = 3$ A

(iii) The current through meter A3 is the phasor sum of the two branch currents. This is found to be 3 A. See Figure 108.

200 This is not a science question. Please modify as follows:
A 240 V, 12 kW single-phase load operates at a power factor of 0.7 lagging.
(a) Calculate the current taken by the load
(b) Obtain graphically or by calculation the kVAr of a power factor improvement capacitor

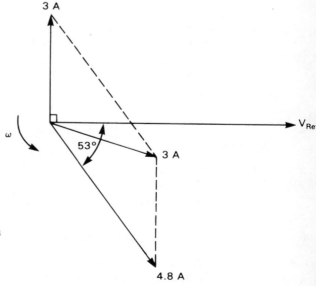

Figure 108

connected in parallel with the load to improve the overall power factor to unity.
(c) Calculate the current taken from the supply after power factor correction.
(d) State TWO advantages of power factor improvement.

CGLI/II/1989

Solution

(a) Since $P = V \times I \times \text{p.f.}$
then $I = P/(V \times \text{p.f.})$
$= 12{,}000/(240 \times 0.7)$
$= 71.43$ A

(b) See Fig. 109. (See page 119 of proofs) Since $\cos \phi$ is the power factor, the phase angle (ϕ) is found to be 45.57. A right-angled power triangle is now constructed. By calculation, since p.f. = Power (P)/Voltamperes(S)

then $S = P/\text{p.f.}$
$= 12/0.7 = 17.14$ kVA
thus $Q = \sqrt{S^2 - P^2}$
$= \sqrt{17.14^2 - 12^2}$
$= 12.24$ kVAr lagging

For unity p.f., a capacitor must inject a leading Q into the circuit, equal to 12.24 kVAr

(c) With the power factor now at unity,
$I = P/V$
$= 12{,}000/240$
$= 50$ A

(d) Two advantages of power factor improvement are:
(i) smaller cables and switchgear
(ii) more economical installation.

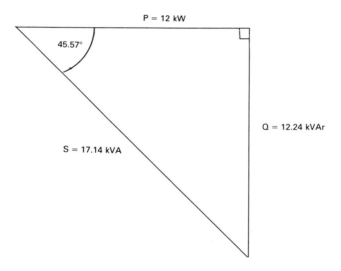

Figure 109 *Power triangle*

Multiple-choice questions

Part I Certificate

Terminology

1. The quantity of electricity (Q) is found by the expression
   (a) $Q = V \times I$
   (b) $Q = P \times t$
   (c) $Q = I \times t$
   (d) $Q = I \times R$

2. The unit of reactance is called the
   (a) volt
   (b) ampere
   (c) watt
   (d) ohm

3. In Figure 110 which is the BS 3093 graphical symbol for a fuse?
   (a) (b) (c) (d)

4. Extra-low voltage a.c. is normally between the range
   (a) 0 V and 50 V
   (b) 50 V and 100 V
   (c) 100 V and 600 V
   (d) 600 V and 1000 V

5. In Figure 111 which is the BS 3939 graphical symbol for a two-pole, one-way switch?
   (a) (b) (c) (d)

6. In Figure 112 which conduit box is called a *tangent tee box*?
   (a) (b) (c) (d)

7. The abbreviation CNE refers to a type of cable having
   (a) combined neutral earth
   (b) cross nylon ethylene
   (c) covering negative earth
   (d) choice neutral envelope

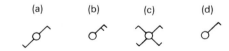

Figure 111

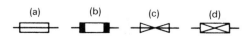

Figure 110

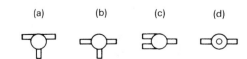

Figure 112

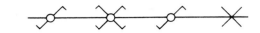

Figure 113

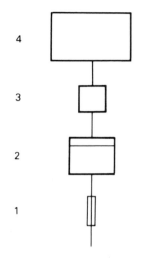

Figure 114

8   The line diagram shown in Figure 113 identifies the control circuit for
    (a)   two-way and intermediate switching
    (b)   one-way and two-way switching
    (c)   two-pole and four-pole switching
    (d)   two-gang and four-gang switching

9   A branch cable connected to a ring final circuit is called the
    (a)   inter-connector
    (b)   strapper
    (c)   extension lead
    (d)   spur

10 In Figure 114 below, items 1 and 2 are the responsibility of the
    (a)   Department of Environment
    (b)   Central Electricity Generating Board
    (c)   Local Electricity Board
    (d)   Electrical Consumer

Health and Safety

11 Immediate action in the case of a person suffering from the effects of toxic fumes is to
    (a)   seek medical assistance
    (b)   open all windows and doors
    (c)   apply artificial respiration
    (d)   remove person from danger area

12 If battery acid is accidentally spilled on the skin, the affected part should immediately be
    (a)   seen by a doctor
    (b)   bandaged with a sterile dressing
    (c)   covered with an antiseptic cream
    (d)   flooded with cold water

13 In Figure 115, which situation of touching the conductors is likely to prove the most serious?
    (a)   (b)   (c)   (d)

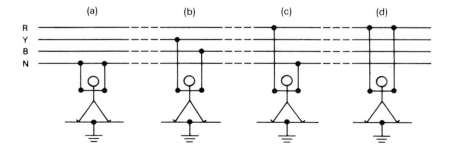

Figure 115

14 Which of the following fire extinguishers is suitable for a live electrical fire?
    (a) carbon dioxide
    (b) water
    (c) foam
    (d) carbon tetrachloride

15 A ladder tied to a working platform should extend at least
    (a) 2.2 m
    (b) 1.8 m
    (c) 1.5 m
    (d) 1.0 m

16 A double-insulated electric power tool should be checked for the following *except*
    (a) correct speed
    (b) supply voltage
    (c) correct fuse
    (d) earthing

17 Which method of earthing is no longer recognized today?
    (a) water pipe
    (b) earth electrode
    (c) cable sheath
    (d) protective multiple earthing

18 In an electric circuit, protection against excess current may be provided by
    (a) earth bonding
    (b) circuit conductors
    (c) suitable size fuse
    (d) earth monitoring

19 With reference to Figure 116, the most suitable ladder ratio is
    (a) 8:3
    (b) 7:4
    (c) 4:1
    (d) 2:1

20 If a patient does not respond to mouth-to-mouth treatment, a first-aider should immediately
    (a) check for broken limbs
    (b) check pupils of eyes, and pulse
    (c) keep the body warm with blankets
    (d) seek medical help

Installation 1

21 In a discharge lamp circuit, the function of the *choke* is to
    (a) improve the power factor
    (b) absorb ultra-violet rays
    (c) reduce the cathode volt drop
    (d) limit the supply current

22 In Figure 117, the potential difference across R_1 is found by
    (a) $V_1 = V_s - V_2 - V_3$
    (b) $V_1 = V_s + V_2 + V_3$
    (c) $V_1 = V_s - V_2 + V_3$
    (d) $V_1 = V_s + V_2 - V_3$

23 In Figure 118, which wattmeter is correctly connected?
    (a) (b) (c) (d)

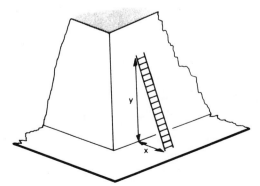

Figure 116

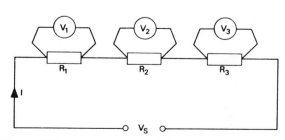

Figure 117

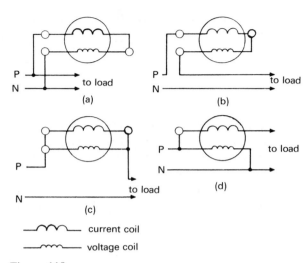

current coil

voltage coil

Figure 118

24 If the charge conveyed in an electric circuit is 300 C, the time taken to pass a current of 3 A is
  (a) 100 s
  (b) 297 s
  (c) 303 s
  (d) 900 s

25 What is the current taken by a resistive load of 400 Ω dissipating 10 kW of heat?
  (a) 5 A
  (b) 10 A
  (c) 15 A
  (d) 25 A

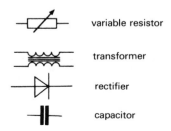

variable resistor

transformer

rectifier

capacitor

Figure 119

26 Which electrical component in Figure 119 operates on the principle of mutual inductance?
  (a) variable resistor
  (b) transformer
  (c) rectifier
  (d) capacitor

27 What is the power consumed by a 10 Ω resistor supplied at 200 V?
  (a) 2 kW
  (b) 4 kW
  (c) 10 kW
  (d) 20 kW

28 Which circuit in Figure 120 will keep the signal lamp on when P is pressed and then released?
  (a) (b) (c) (d)

29 The current taken by a 60 W, 240 V tungsten lamp is
  (a) 100 mA
  (b) 220 mA
  (c) 250 mA
  (d) 400 mA

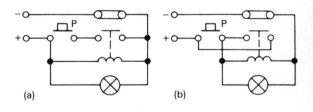

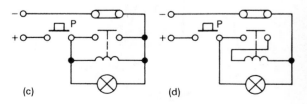

Figure 120

Multiple Choice: Part I

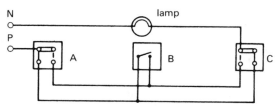

Figure 121

30 The value of a sine wave corresponding to 0.707 of maximum value is called the
   (a) peak value
   (b) average value
   (c) instantaneous value
   (d) root mean square value

31 When switch B is closed, the lamp in Figure 121
   (a) can only be switched off from A
   (b) cannot be switched off from A or B
   (c) immediately becomes dim
   (d) short-circuits

32 With reference to Figure 122, which factor below does not alter the strength of induced e.m.f.?
   (a) magnetic field density
   (b) effective conductor length
   (c) conductor velocity
   (d) permeability

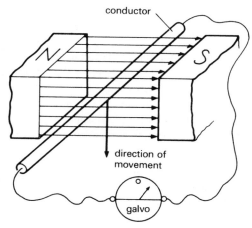

Figure 122

33 The term *overcurrent* may be used to describe an overload current or
   (a) short-circuit current
   (b) earth-fault current
   (c) earth-neutral current
   (d) harmonic current

34 What is the estimated short-circuit current in part of a system in which the 240 V supply drops to 235 V when the load current is 25 A?
   (a) 1200 A
   (b) 1000 A
   (c) 500 A
   (d) 125 A

35 Balancing single-phase loads on three-phase, four wire systems is to ensure that
   (a) line voltages are all equal
   (b) minimal neutral current flows
   (c) star point is maintained at all times
   (d) circuit fuses operate efficiently

36 In Figure 123, the system would be ideally balanced if the yellow phase load X carried
   (a) 30 A
   (b) 50 A
   (c) 64 A
   (d) 82 A

37 Dividing discharge lamps over three phases is a method of overcoming
   (a) poor power factor
   (b) uneven light distribution
   (c) stroboscopic effects
   (d) wiring difficulties

38 To obtain an equivalent circuit resistance of 20 Ω (Figure 124) the following switches require closing
   (a) S1, S2, S3, S4
   (b) S2, S3, S4, S5
   (c) S3, S4, S5, S6
   (d) S4, S5, S6, S7

39 Gas filling of a tungsten filament lamp is carried out to
   (a) evacuate bulb impurities
   (b) increase internal bulb wall pressure
   (c) reduce ultra-violet radiation
   (d) reduce filament evaporation

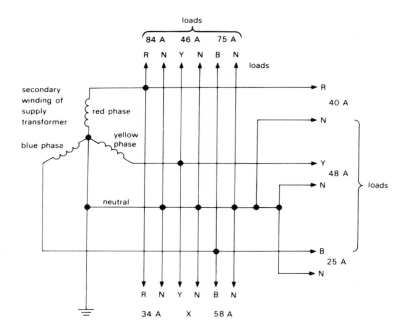

Figure 123

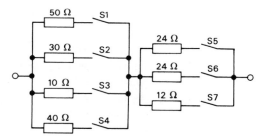

Figure 124

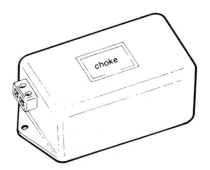

Figure 125

40 The power factor of an a.c. circuit is given by the ratio

(a) $\dfrac{\text{power output}}{\text{power input}}$

(b) $\dfrac{\text{wattless power}}{\text{reactive power}}$

(c) $\dfrac{\text{true power}}{\text{apparent power}}$

(d) $\dfrac{\text{average power}}{\text{maximum power}}$

41 A power factor correction capacitor is used in discharge lamp control gear for the purpose of
(a) reducing high-voltage surges
(b) reducing harmonic distortion
(c) stabilizing the light output
(d) neutralizing the effects of inductance

42 The discharge lamp ballast shown in Figure 125 limits current flow because of its
(a) high-voltage surge
(b) high capacitance
(c) inductive reactance
(d) resistance property

Multiple Choice: Part I

43 A double-wound transformer delivers 500 V to a load taking 500 kVA. What current is taken by the load?
 (a) 100 A
 (b) 500 A
 (c) 1000 A
 (d) 5000 A

44 When S1 and S2 are closed (Figure 126), which position is suitable for connecting a signal lamp to show that the bell has been silenced?
 (a) circle 1
 (b) circle 2
 (c) circle 3
 (d) circle 4

45 The name of the device marked A in Figure 126 is called
 (a) an alarm switch
 (b) a diversion relay
 (c) a contact breaker
 (d) an emergency push

46 The electrolyte in a lead-acid cell is
 (a) sal-ammoniac solution
 (b) hydrochloric acid
 (c) dilute sulphuric acid
 (d) potassium hydroxide

47 The specific gravity of an alkaline cell during charge
 (a) increases with testing
 (b) decreases with volt drop
 (c) remains constant
 (d) varies with time

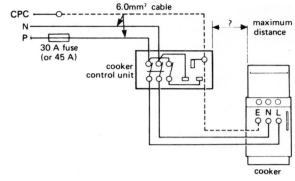

Figure 127

48 In Figure 127, the control switch should be placed no further away than
 (a) 1 m
 (b) 2 m
 (c) 3 m
 (d) 4 m

49 The minimum distance a 13 A socket-outlet can be wired to a shower cubicle in a bedroom is
 (a) 1.0 m
 (b) 2.5 m
 (c) 3.6 m
 (d) 4.0 m

50 The device shown in Figure 128 is a Simmer energy regulator used for controlling
 (a) electronic amplifiers
 (b) resistance heating elements
 (c) radio capacitors
 (d) inductive ballast coils

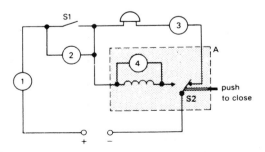

Figure 126

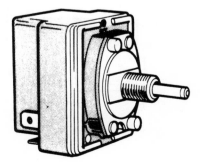

Figure 128

Materials

51 Which of the following conductors has the lowest resistivity value?
   (a) hard-drawn aluminium
   (b) annealed copper
   (c) rolled brass
   (d) tungsten

52 A suitable material to use as a connector block inside an electric storage heater is
   (a) Paxolin   (c) PVC
   (b) Bakelite   (d) porcelain

53 Thermoplastic PVC cables should not be handled where the temperature is likely to be
   (a) less than 0°C
   (b) between 1°C and 10°C
   (c) between 10°C and 15°C
   (d) between 15°C and 20°C

54 Electrical conductor resistance depends upon
   (a) insulation thickness
   (b) length
   (c) shape
   (d) type of sheath

55 Which of the following is a *ferrous* metal?
   (a) iron   (c) copper
   (b) aluminium   (d) lead

56 In Figure 129, the moving part of the bell attracted towards the coil is called the
   (a) hammer   (c) keeper
   (b) armature   (d) spring

57 Contact between brass and aluminium is likely to cause
   (a) ionization
   (b) condensation
   (c) sulphuration
   (d) corrosion

58 A suitable material to use when making good a surrounding hole in a wall (Figure 130) is
   (a) compound
   (b) wood
   (c) cement
   (d) sacking

59 A conductor 100 metres long has an insulation resistance of 50 megohms. What will be its insulation resistance for a length of 500 m?
   (a) 250 megohm
   (b) 120 megohm
   (c) 25 megohm
   (d) 10 megohm

60 Which of the following materials has a negative temperature coefficient of resistance?
   (a) carbon
   (b) brass
   (c) silver
   (d) copper

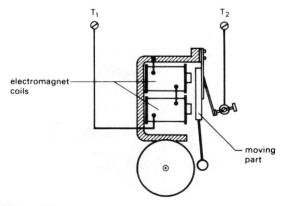

Figure 129

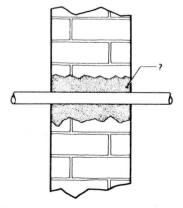

Figure 130

Multiple Choice: Part I

Tools and equipment

61 The conduit threading tool part shown in Figure 131 is called a
   (a) die
   (b) stock
   (c) guide
   (d) sleeve

62 Which tools would you exclude from selection when marking and drilling a steel plate for glands?
   (a) scriber and hole cutters
   (b) footprints and pliers
   (c) hammer and round file
   (d) centre punch and steel ruler

63 What is the length of XY in Figure 132?
   (a) 80.0 mm
   (b) 55.0 mm
   (c) 52.0 mm
   (d) 42.5 mm

Figure 131

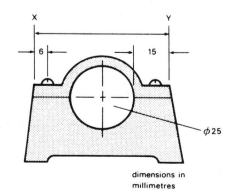

Figure 132

Figure 133

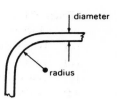

Figure 134

64 Figure 133 shows a
   (a) wing bolt
   (b) toggle bolt
   (c) butterfly clip
   (d) ceiling clip

65 A plumb-bob and line is used for
   (a) cable jointing work
   (b) plumbing work
   (c) vertical alignment work
   (d) drawing-in cables through conduit

66 Which of the following is *not* a tool?
   (a) a tap wrench
   (b) an electric motor
   (c) a hammer drill
   (d) a laminate trimmer

67 Which tool listed below is likely to suffer a 'mushroom-head' if used constantly?
   (a) a cold chisel
   (b) a pein hammer
   (c) a screwdriver
   (d) footprints

68 The distance saddle shown in Figure 132 is used on surfaces subjected to
   (a) extreme cold temperatures
   (b) heavy condensation
   (c) constant vibration
   (d) dust and grease

69 The inner radius of a right-angled steel conduit bend (Figure 134) must not be *less than*
   (a) 2.5 times the outside diameter of the conduit
   (b) 4.0 times the outside diameter of the conduit
   (c) 5.5 times the outside diameter of the conduit
   (d) 6.0 times the outside diameter of the conduit

70 Which of the following soldering fluxes should be used when making an electrical connection?
(a) Baker's fluid
(b) resin
(c) zinc chloride
(d) ammonium chloride

Installation 2

71 Which of the following conduit systems is the most suitable for a flame-proof installation?
(a) solid-drawn metal conduit
(b) flexible metallic conduit
(c) heavy-gauge welded seam conduit
(d) PVC conduit

72 Item 3, shown in Figure 135, is called a
(a) fire alarm push button
(b) luminous push button
(c) key operated switch
(d) heating appliance

73 The type of switch shown in Figure 136 is called a
(a) double-pole linked switch
(b) single-phase contact breaker
(c) change-over switch
(d) switch splitter

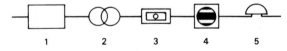

Figure 135

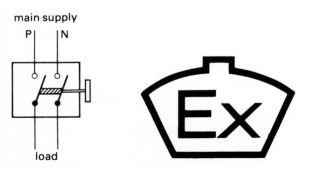

Figure 136   Figure 137

74 Materials delivered to a site should be checked against the delivery note and
(a) the job sheet
(b) the time sheet
(c) the original order
(d) the variation order

75 In the IEE Wiring Regulations, the correction factor for thermal insulation completely surrounding a cable is
(a) 1.00
(b) 0.75
(c) 0.5
(d) 0.1

76 The symbol in Figure 137 is used to denote
(a) externally used apparatus
(b) explosive equipment
(c) exhibition equipment
(d) self-excited machines

77 Protection against direct contact of 'live' parts can be achieved by any of the following *except* for
(a) insulation
(b) barriers
(c) obstacles
(d) bonding

78 Fusing factor is the ratio
(a) $\dfrac{\text{line current}}{\text{phase current}}$
(b) $\dfrac{\text{prospective current}}{\text{fusing current}}$
(c) $\dfrac{\text{close protection}}{\text{coarse protection}}$
(d) $\dfrac{\text{fusing current}}{\text{current rating}}$

79 Lampholders for filament lamps should not be used above
(a) 250 V
(b) 240 V
(c) 230 V
(d) 110 V

Multiple Choice: Part I

80 The correction factor given in the IEE Wiring Regulations for thermal insulation is
  (a) Ca
  (b) Ci
  (c) Cg
  (d) Ct

81 An example of an *extraneous conductive part* is
  (a) a metal sink
  (b) an electrical conductor
  (c) a metalclad fuseboard
  (d) an electric kettle element

82 In vertical trunking systems containing cables, fire barriers are required between floors at a distance *not exceeding*
  (a) 8 m
  (b) 5 m
  (c) 4 m
  (d) 2 m

83 Which measurement in Figure 138 shows the size of conduit?
  (a) W
  (b) X
  (c) Y
  (d) Z

84 The test on a residual current device by a consumer should be carried out
  (a) yearly
  (b) quarterly
  (c) monthly
  (d) weekly

85 A 30 m length of twin PVC insulated cable is rated at 24 A with a 10 mV drop/A/m. What is its actual volt drop when it carries a current of 20 A?
  (a) 4.8 V   (c) 6.0 V
  (b) 5.0 V   (d) 7.5 V

86 The current demand of a cooking appliance is based on the first 10 A of total rated current plus 30% of the remainder total rated current, plus 5 A if the control unit incorporates a socket-outlet. For a 6 kW/240 V cooker based on the information above, the current demand is
  (a) 30.0 A   (c) 19.5 A
  (b) 25.2 A   (d) 15.0 A

87 A crampet is a fitting used on
  (a) conduit work
  (b) trunking work
  (c) MIMS cable wiring
  (d) duct system wiring

88 In Figure 139, the depth for drilling a hole in the wooden joist should be
  (a) 100 mm
  (b) 75 mm
  (c) 63 mm
  (d) 50 mm

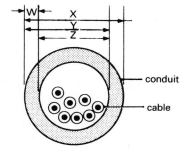

Figure 138

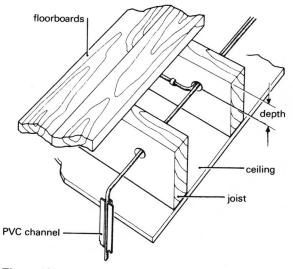

Figure 139

89 The IEE Wiring Regulations exclude from its scope *all* the following *except*
(a) ship installations
(b) caravan installations
(c) offshore installations
(d) aircraft installations

90 The circuit classification for a fire alarm system is
(a) category 0
(b) category 1
(c) category 2
(d) category 3

Testing and inspection

91 With the switch closed, the appliance shown in Figure 140 is being tested for
(a) polarity
(b) insulation resistance
(c) earthing effectiveness
(d) element continuity

92 In testing an installation for insulation resistance, large sections of the wiring may be divided into groups of outlets, each containing not less than
(a) 200 outlets
(b) 150 outlets
(c) 100 outlets
(d) 50 outlets

93 The type of instrument used for testing the effectiveness of the earth fault loop is called
(a) an impedance tester
(b) a metal detector
(c) a soil resistivity tester
(d) a tong tester

94 The a.c. test voltage being applied to test the electrical continuity of a protective earth conductor *must not exceed*
(a) 250 V
(b) 100 V
(c) 50 V
(d) 25 V

95 Which row is correct for the colour identification of a three-phase, four-wire flexible cable?

	live	neutral	earth
(a)	red	black	green and yellow
(b)	red	blue	green
(c)	brown	blue	green and yellow
(d)	brown	black	green

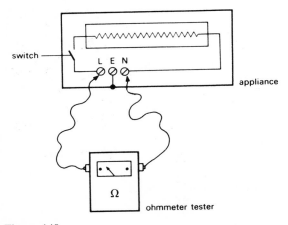

Figure 140

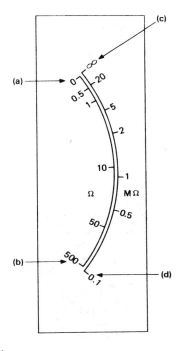

Figure 141

Multiple Choice: Part I

96 In Figure 141, a 'dead-short' on the continuity scale of the tester would be indicated by
(a) 0
(b) 500
(c) ∞
(d) 0.1

97 The minimum insulation resistance test figure for a completed installation is
(a) 50 megohm
(b) 10 megohm
(c) 1 megohm
(d) ½ megohm

98 Good earth continuity implies
(a) high resistance
(b) low resistance
(c) open circuit
(d) low conductivity

99 Which one of the following instruments is used for making a resistance test?
(a) wattmeter
(b) voltmeter
(c) ammeter
(d) multimeter

100 It is recommended to test a temporary installation after a period of
(a) 3 months
(b) 6 months
(c) 2 years
(d) 5 years

Part II Certificate

Communications and industrial studies

101 *All apparatus and conductors shall be sufficient in size and power for the work they are called upon to do.* This is a requirement from the
   (a) Building Regulations
   (b) Electrical Appliance Regulations
   (c) Health and Safety at Work etc. Act
   (d) Electricity (Factories Act) Special Regulations

102 The main purpose of a bar chart used in electrical work is to show
   (a) work activities for men on site
   (b) work grades of craftsmen for specific assignments
   (c) estimated project dates
   (d) estimated costs on site

103 A contract document is an assurance that
   (a) labour on site can be specially programmed
   (b) the work and the payment will be carried out
   (c) good site management will be guaranteed
   (d) the building owner and contractor will work together

104 The body responsible in electrical contracting for determining wages and working conditions is the
   (a) Joint Industry Board
   (b) Employers' Association
   (c) Advisory, Conciliation and Arbitration Service
   (d) Wages Council

105 A bill of quantities is a
   (a) method of tendering
   (b) list of extra work
   (c) schedule of rates
   (d) variation order

106 An architect's representative on site is called a
   (a) clerk of works
   (b) main contractor
   (c) quantity surveyor
   (d) consulting engineer

107 A drawing marked 'as fitted' is one that shows
   (a) block and layout diagrams
   (b) records of work done
   (c) revised structural alterations
   (d) all sub-contract work

108 The term 'specification' used in electrical contracting implies
   (a) manufacturer's design data
   (b) erection and wiring information only
   (c) general and specific requirements for carrying out work
   (d) statutory requirements for a given contract

109 The BS 3939 symbol shown in Figure 142 is called
   (a) telephone point
   (b) emergency lighting point
   (c) indicator panel
   (d) automatic fire detector

Figure 142

110 The diagram in Figure 143 is known as a
   (a) block diagram
   (b) layout diagram
   (c) schematic diagram
   (d) line diagram

Health and safety

111 A written authorization to carry out work on or about live equipment is called
   (a) a completion certificate
   (b) an inspection certificate
   (c) a permit to work document
   (d) a variation order

112 A plug and socket can only be used for (1) *emergency switching* and (2) *isolation*.
   (a) Both (1) and (2) are true
   (b) Both (1) and (2) are false
   (c) Only (1) is true
   (d) Only (2) is true

113 To prevent danger, the Factories Act Regulations require all motor driven machines to have means of
   (a) identification of make
   (b) lifting in and out of position
   (c) stopping and starting facilities
   (d) reversal of direction

114 In British Standard BS 5345 Part 1: 1976, an area in which a dangerous atmosphere is continuously present in normal operation is called
   (a) zone 0
   (b) zone 1
   (c) zone 2
   (d) zone 3

115 The minimum size of supplementary bonding conductor permitted in a bathroom where no mechanical protection is provided is
   (a) 4.0 mm^2
   (b) 2.5 mm^2
   (c) 1.5 mm^2
   (d) 1.0 mm^2

116 To avoid contact near live electrical conductors and apparatus it is necessary to provide
   (a) printed instructions for all unskilled labour
   (b) temporary earthing facilities
   (c) insulated screens in the area of work
   (d) earth leakage protection on supplies above 11 kV

117 In Figure 144, the IEE Wiring Regulations require the minimum height of span to be
   (a) 5.8 m
   (b) 4.8 m
   (c) 4.2 m
   (d) 3.5 m

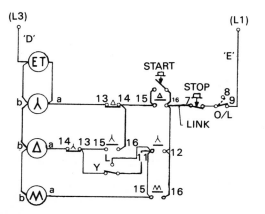

Figure 143

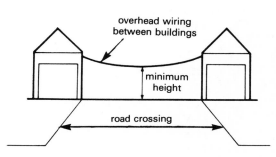

Figure 144

118 To reduce the risk of an electric shock, a capacitor bank is fitted with
(a) an automatic cut-out device
(b) a discharge resistor
(c) a heat sink
(d) an earth-leakage trip

119 Cables buried underground between buildings should be installed (1) at sufficient depth to avoid damage, and (2) be marked with cable covers or suitable marking tape.
(a) both statements are true
(b) both statements are false
(c) only statement (1) is true
(d) only statement (2) is true

120 A fuse may provide both (1) excess current protection, and (2) earth-leakage protection if the earth impedance is low enough.
(a) both statements are true
(b) both statements are false
(c) only statement (1) is true
(d) only statement (2) is true

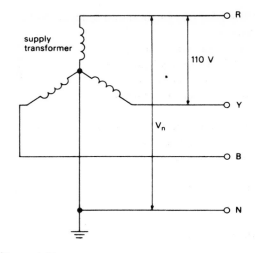

Figure 145

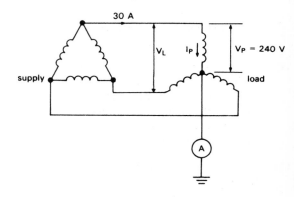

Figure 146

Distribution

121 In Figure 145, the line to neutral voltage (V_n) is approximately
(a) 110 V
(b) 84 V
(c) 64 V
(d) 55 V

122 Which of the following is not recognized as a standard a.c. supply voltage in the UK?
(a) 11 000 V
(b) 650 V
(c) 415 V
(d) 240 V

123 In Figure 146 the load phase current is
(a) 90 A
(b) 45 A
(c) 30 A
(d) 20 A

124 In Figure 37 the load line voltage is
(a) 415 V
(b) 320 V
(c) 240 V
(d) 110 V

125 In Figure 146 if all line conductors carried 30 A, the neutral ammeter would read
(a) 90 A
(b) 30 A
(c) 10 A
(d) 0 A

126 Opening one phase of a three-phase, three-wire balanced system would
(a) increase the current flow in the two healthy phases
(b) decrease the current flow in the two healthy phases
(c) make no difference to the current flow in the healthy phases
(d) cause no current flow in the healthy phases

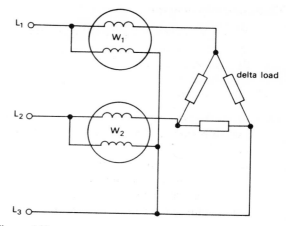

Figure 148

127 Which is correct for a delta connected system
(a) $I_L = I_p$ and $V_L = V_p$
(b) $I_L = \sqrt{3} I_p$ and $V_L = \sqrt{3} V_p$
(c) $I_L = \sqrt{3} I_p$ and $V_L = V_p$
(d) $I_L = I_p$ and $V_L = \sqrt{3} V_p$
where I_p is the phase current
I_L is the line current
V_p is the phase voltage
V_L is the line voltage

128 Which diagram (see Figure 147) illustrates a TN-C-S system?

129 In Figure 148 the total power is found by one of the following
(a) $W_1 + W_2$
(b) $W_1 - W_2$
(c) $W_1 \div W_2$
(d) $W_1 \times W_2$

130 In Figure 149 the power factor is
(a) 0.8
(b) 0.7
(c) 0.6
(d) 0.5

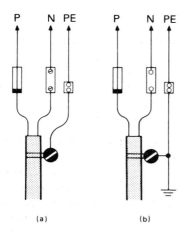
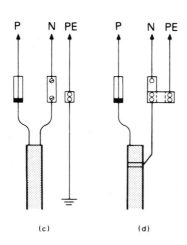

Figure 147

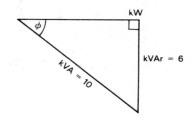

Figure 149

Consumer's switchgear

131 The device shown in Figure 150 is called a
   (a) linked switch
   (b) contact breaker
   (c) fused switch
   (d) switchfuse

132 *Prospective current* and *pre-arcing time* are two axes of reference which show the characteristics of
   (a) excess current protective devices
   (b) transformers operating in parallel
   (c) alternators operating in parallel
   (d) solid-state semiconductor devices

133 Ideal discrimination between two fuses of different size is when
   (a) the fuse nearest the fault provides limited protection
   (b) both fuses operate together
   (c) the fuse nearest the fault operates first
   (d) the fuse nearest the fault operates last

134 Where a rising main busbar system is used in a tall building, arrangements need to be made for
   (a) busbars expanding and contracting
   (b) busbars having no movement
   (c) protecting the busbars against lightning
   (d) protecting the busbars against corrosion

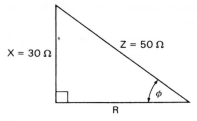

Figure 151

135 In Figure 151 the resistance and power factor respectively are found to be
   (a) 45 Ω and 0.75
   (b) 40 Ω and 0.80
   (c) 35 Ω and 0.85
   (d) 31 Ω and 0.90

136 The device shown in Figure 152 only operates on the occurrence of
   (a) a high impedance
   (b) a short-circuit
   (c) an earth-leakage
   (d) an open circuit

137 The method of bonding items of metalwork together is to
   (a) avoid the flow of neutral currents
   (b) ensure that a common potential exists
   (c) safeguard against short circuit faults
   (d) prevent corrosion of the earthing system

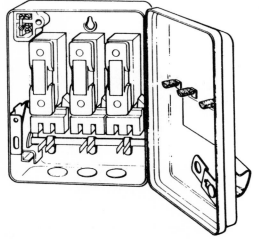

Figure 150

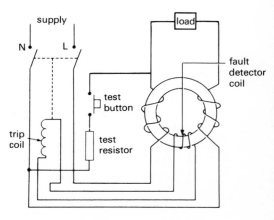

Figure 152

Multiple Choice: Part II

138 Warning notices, clearly visible to indicate the maximum voltage present are required on all equipment and enclosures operating above
  (a) 600 V
  (b) 500 V
  (c) 415 V
  (d) 250 V

139 The term used as an allowance for calculating the size of circuit conductors and switchgear of circuits other than final circuits is called
  (a) load factor
  (b) growth factor
  (c) diversity factor
  (d) rating factor

140 Which of the following *cannot* be used as a means of isolation in a circuit?
  (a) fireman's switch
  (b) plug and socket
  (c) thyristor
  (d) fuse link

Wiring systems

141 In which of the following cables is the insulant magnesium oxide powder used?
  (a) PVC armoured cable
  (b) PVC/PVC cables
  (c) MIMS cables
  (d) XLPE cables

142 The main function of the compression ring inside an MIMS cable gland is to provide
  (a) earth continuity
  (b) mechanical protection
  (c) conductor alignment
  (d) protection against corrosion

143 MIMS cable is a suitable wiring system for a
  (a) farm installation
  (b) boiler installation
  (c) temporary installation
  (d) domestic installation

144 The link shown in Figure 153 at the end of the metal trunking provides
  (a) earth continuity
  (b) lightning protection
  (c) mechanical fixing
  (d) external connection

145 All cables of a.c. circuits should be contained in the same metal conduit to avoid
  (a) eddy current flow
  (b) unnecessary circuit loading
  (c) unnecessary circuit wiring
  (d) excessive volt drop

146 In earth concentric wiring, the outer sheath acts both as a circuit protective conductor and
  (a) lightning conductor
  (b) neutral conductor
  (c) phase conductor
  (d) line conductor

147 To overcome the problem of moisture in the end of an MIMS cable, the termination is
  (a) re-made after inspection
  (b) tested with an ohmmeter to drive the moisture away
  (c) shaken violently and then re-made
  (d) dried out with a heat source and then re-made

148 In a flame-proof installation, a conduit stopper box should be fitted
  (a) only at termination points
  (b) every five metres of run
  (c) on the side remote from a danger area
  (d) on the side classified a danger area

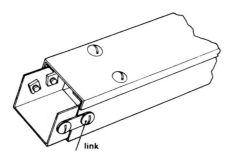

Figure 153

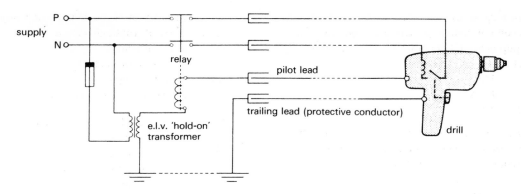

Figure 154

149 The method of protection shown in Figure 154 is often called
(a) earth monitoring
(b) grounding
(c) double earthing
(d) double insulation

150 Before selecting a cable for a final circuit, two conditions have to be met, namely (1) the cable must be capable of carrying the design current of the circuit, and (2) the cable must not have a volt drop exceeding ± 6% of the declared voltage
(a) both (1) and (2) are true
(b) both (1) and (2) are false
(c) only (1) is true
(d) only (2) is true

Installation of motors

151 Which of the following can be described as a split-phase motor?
(a) a shunt motor
(b) a universal motor
(c) a compound motor
(d) a capacitor-start motor

152 The type of rotor shown in Figure 155 is called
(a) slip-ring
(b) armature
(c) wound
(d) cage

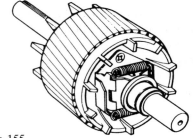

Figure 155

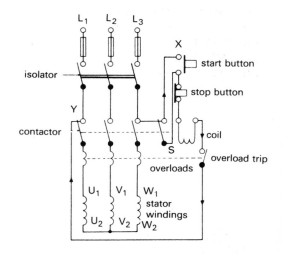

Figure 156

153 To reverse the direction of the motor shown in Figure 156, which of the following connections need to be changed?
(a) U_1 and U_2
(b) L_1 and L_2
(c) X and Y
(d) S and X

154 In an induction motor, per unit slip (s) is found by the expression
(a) $s = (n_r - n_s) \times n_s$
(b) $s = (n_s - n_r) \div n_s$
(c) $s = (n_r + n_s) - n_r$
(d) $s = (n_s + n_r) + n_r$
where n_s is the synchronous speed
n_r is the rotor speed

155 In a d.c. motor, the back e.m.f. (E) is given by the expression
(a) $E = V - I_a R_a$
(b) $E = V + I_a R_a$
(c) $E = V \div I_a R_a$
(d) $E = V \times I_a R_a$
where V is the supply voltage
I_a is the armature current
R_a is the armature resistance

156 In the d.c. motor shown in Figure 157 the effect of increasing field resistance is to
(a) decrease the speed of the motor
(b) increase the speed of the motor
(c) limit the motor's starting current
(d) reduce commutator sparking

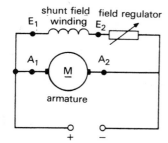
Figure 157

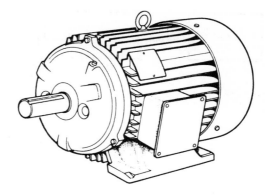
Figure 158

157 The motor shown in Figure 158 with the identification Ex N is used in
(a) damp locations
(b) hazardous locations
(c) underwater locations
(d) non-hazardous locations

158 The method of connecting capacitors in delta across the lines of a three-phase induction motor is to
(a) limit the inrush of current to the stator
(b) improve the power factor
(c) increase the synchronous speed
(d) make it run at a constant speed

159 All the factors below may alter the strength of an induced e.m.f. in a rotating conductor situated in a magnetic field *except*
(a) magnetic flux
(b) conductor length
(c) conductor velocity
(d) potential difference

160 To reverse the direction of a d.c. series motor it is necessary to interchange (1) either armature winding connections or field winding connections, and (2) supply connections
(a) both statements (1) and (2) are correct
(b) both statements (1) and (2) are incorrect
(c) only statement (1) is correct
(d) only statement (2) is correct

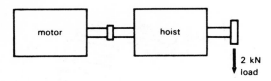

Figure 159

161 With a per unit slip of 0.07, the rotor speed of a four-pole cage induction motor fed from a 50 Hz supply is
(a) 25.65 rev/s
(b) 24.48 rev/s
(c) 23.25 rev/s
(d) 22.11 rev/s

162 Figure 159 shows a motor driving a hoist which is used to raise a load through a vertical distance of 60 m in 80 s. What is the motor's rated output if the hoist is 60% efficient and the load 2000 N?
(a) 8.3 kW
(b) 6.6 kW
(c) 4.5 kW
(d) 2.5 kW

163 What is the efficiency of a 250 V d.c. motor taking a current of 16.8 A if its rated output is 3730 W?
(a) 88.8%
(b) 70.5%
(c) 66.7%
(d) 50.4%

164 Where the motor requires to be coupled to a driven machine, it is important to ensure that both shafts are in line. One method of doing this is to
(a) measure the distance from the floor to the centre of each shaft
(b) insert a feeler gauge between the flange faces of both shafts
(c) bolt both shafts together and apply equal pressure when tightening up
(d) run a metal wire from one machine to the other, in parallel to both shafts

165 The power dissipated in each star-connected winding of a three-phase motor is found by the expression $P = V_p I_p \cos \phi$. What is the total power if $V_p = 240$ V, $I_p = 2.25$ A and $\cos \phi = 0.5$ lagging?
(a) 810 W
(b) 540 W
(c) 270 W
(d) 110 W

166 The use of a star-delta starter in the operation of a six terminal induction motor reduces the starting voltage across each phase winding by approximately
(a) 70% of the line value
(b) 62% of the line value
(c) 58% of the line value
(d) 43% of the line value

167 In the graph shown in Figure 160, the starting torque is indicated by which number?
(a) 1
(b) 2
(c) 3
(d) 4

168 Which of the following motors has copper shading rings for delaying magnetic flux in the stator?
(a) repulsion motor
(b) shaded pole motor
(c) universal motor
(d) compound motor

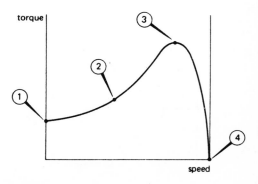

Figure 160

169 A three-phase induction motor refuses to start, the following are likely causes *except*
 (a) fuse blown in the line
 (b) open circuit in a stator winding
 (c) open circuit between stator and rotor
 (d) open circuit in rotor winding

170 Blackening of a d.c. motor's commutator is brought about by the following *except*
 (a) brushes sticking
 (b) short-circuiting armature field coils
 (c) compoles wrongly connected
 (d) open circuit field winding

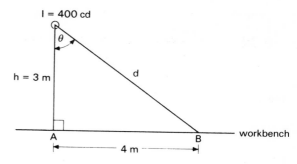

Figure 161

Installation of lighting

171 The measure of light falling on a surface is called
 (a) luminance
 (b) luminous flux
 (c) luminous intensity
 (d) illuminance

172 The ratio $\dfrac{\text{useful flux}}{\text{total emitted flux}}$ is called
 (a) coefficient of utilization
 (b) maintenance factor
 (c) diversity factor
 (d) room index

173 The unit of luminous intensity is the
 (a) lux
 (b) lumen
 (c) candela
 (d) nit

174 With reference to Figure 161, the illuminance at point A is approximately
 (a) 133 lux
 (b) 100 lux
 (c) 44 lux
 (d) 7 lux

175 With reference to Figure 161, the illuminance at point B is
 (a) 44.4 lux
 (b) 16.0 lux
 (c) 9.6 lux
 (d) 3.1 lux

176 Lamp efficacy is given by the ratio
 (a) $\dfrac{\text{lumens}}{\text{watt}}$
 (b) $\dfrac{\text{candelas}}{\text{metres}}$
 (c) $\dfrac{\text{lux}}{4\pi r^2}$
 (d) $\dfrac{\text{luminous intensity}}{\text{luminous flux}}$

177 In some discharge lamp installations, rotating machinery may appear to be stationary because of
 (a) poor circuit power factor
 (b) mixed lamp colouring
 (c) stroboscopic effect
 (d) unbalanced circuits

178 Where discharge lamp circuit details are not provided the voltamperes required may be multiplied by a factor of
 (a) 3.5
 (b) 2.0
 (c) 1.8
 (d) 1.5

glow-type starter

Figure 162

179 The device shown in Figure 162 is used for starting
 (a) high-pressure mercury vapour lamps
 (b) low-pressure mercury vapour lamps
 (c) high-pressure sodium vapour lamps
 (d) low-pressure sodium vapour lamps

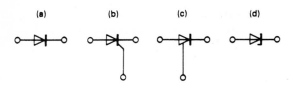

Figure 163

180 The maximum voltage to earth for a transformer supplying a discharge lamp installation is
(a) 10 kV
(b) 5 kV
(c) 1 kV
(d) 415 V

Semiconductor devices

181 Which of the semiconductor symbols shown in Figure 163 is called a zener diode?
(a) (b) (c) (d)

182 Which circuit shown in Figure 164 will both lamps light?
(a) (b) (c) (d)

183 When reverse biased, a p-n junction diode has a
(a) low resistance
(b) zero resistance
(c) high resistance
(d) negative resistance

184 The unidirectional pulse produced by the bridge rectifier network shown in Figure 165 is

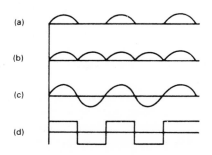

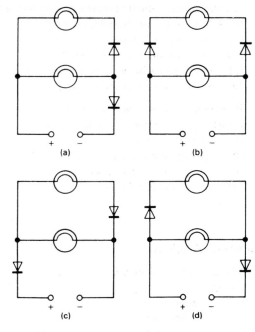

Figure 164

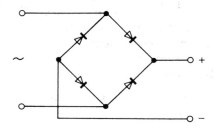

Figure 165

185 In Figure 166, the lamp will light only when
(a) S1 is closed and S2 is open
(b) S2 is closed and S1 is open
(c) S1 and S2 are both closed
(d) S1 and S2 are both open

186 A mobile vacancy amongst valence electrons in a semiconductor material is called
(a) air gap
(b) hole
(c) covalent bond
(d) ion

Multiple Choice: Part II

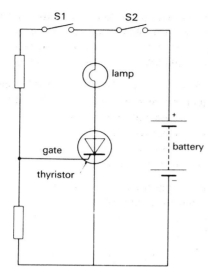

Figure 166

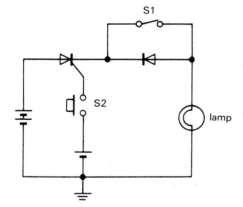

Figure 167

187 The electrode normally used to control a thyristor device is called the
(a) gate
(b) anode
(c) cathode
(d) base

188 In order to ensure the reliable operation of a diode or thyristor against voltage transients, their circuits should be fitted with
(a) linear resistors
(b) surge suppressors
(c) lightning conductors
(d) earth-leakage trips

189 The lamp shown in Figure 167 will only light when
(a) S1 is closed and S2 is pressed
(b) S1 is open and S2 is pressed
(c) S1 and S2 are both open
(d) S1 is closed and S2 is open

190 In order to improve the output voltage pulse of a full-wave rectifier circuit, use is made of a
(a) filtering circuit
(b) feedback circuit
(c) barrier potential
(d) line zener diode

Testing and measurement

191 Testing a wiring installation for insulation resistance is to ensure that
(a) all conductors have high ohmic values
(b) all outlet points are earthed
(c) excess leakage-current does not escape to earth
(d) live and neutral conductors are continuous

192 In the IEE Wiring Regulations an inspection certificate should always accompany, and be attached to, a completion certificate. This requirement is
(a) true for all installations
(b) untrue for all installations
(c) only true for new installations
(d) only true for major installations

193 In Figure 168, the supply phase conductor feeding the third row of fuses should be identified as the
(a) red phase
(b) brown phase
(c) blue phase
(d) yellow phase

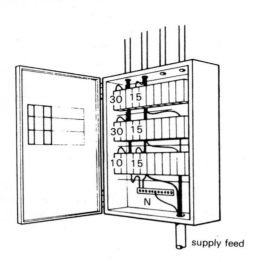

Figure 168

194 The insulation resistance of a cable is 200 MΩ per km. The ohmic value for 250 m is
   (a) 800 MΩ   (c) 250 MΩ
   (b) 500 MΩ   (d) 50 MΩ

195 In Figure 169, the distances D1 and D2 should be equal and the resistance areas of X and Y
   (a) compared with each other
   (b) reduced by one-half ohms
   (c) should not overlap
   (d) made to overlap

196 The voltage used to test continuity of a protective conductor should not exceed
   (a) 250 V   (c) 64 V
   (b) 110 V   (d) 50 V

197 To extend the range of a voltmeter a multiplier is used having
   (a) low ohmic resistance
   (b) high ohmic resistance
   (c) low ohmic reactance
   (d) high ohmic reactance

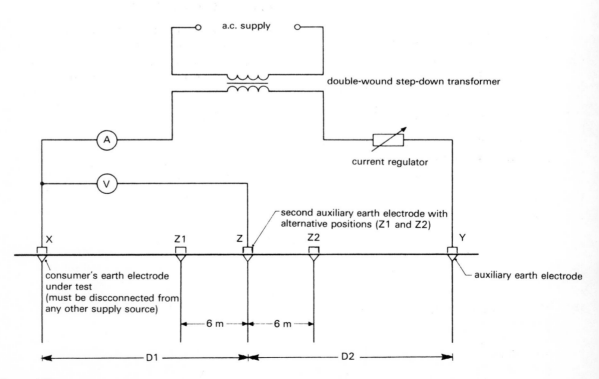

Figure 169

Multiple Choice: Part II

Figure 170

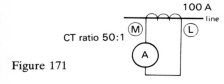

Figure 171

198 The device shown in Figure 170 is called
(a) a frequency meter
(b) a tong test ammeter
(c) an instrument current transformer
(d) an instrument voltage transformer

199 In Figure 171, if the line current is 100 A, the ammeter would read
(a) 5 A
(b) 4 A
(c) 3 A
(d) 2 A

200 The results of individual insulation resistance tests carried out on a completed installation are 100 megohms, 50 megohms, 1 megohm and 0.2 megohms respectively. An overall test reading would indicate
(a) over 100 megohms
(b) between 50 and 100 megohms
(c) between 1 megohm and 50 megohms
(d) less than 0.2 megohms

Revision exercise one

201 A 3 kW, 240 V immersion heater may be protected by a BS 1361 rated at
   (a) 30 A
   (b) 20 A
   (c) 15 A
   (d) 10 A

202 In Figure 172 which is the BS 3939 graphical symbol for a telephone point?
   (a) (b) (c) (d)

203 A kilowatt-hour meter records
   (a) power
   (b) frequency
   (c) charge
   (d) energy

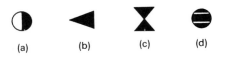

Figure 172

204 The unit of inductance is called the
   (a) henry
   (b) farad
   (c) coulomb
   (d) tesla

205 An instrument using *lux* as its unit of measurement is called
   (a) a magnetic flux meter
   (b) an energy meter
   (c) a frequency meter
   (d) a light meter

206 A reamer tool is used for
   (a) lifting floor grids
   (b) drilling holes in metalwork
   (c) marking out lines on metal plates
   (d) removing burrs from conduit

207 The use of a capital M preceding a unit or unit symbol denotes
   (a) metre
   (b) micro
   (c) mega
   (d) milli

208 The number of units of electricity read by the dials shown in Figure 173 are
   (a) 95 794
   (b) 95 695
   (c) 94 795
   (d) 94 694

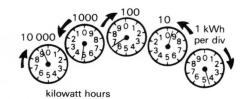

Figure 173

Revision exercise one

209 A *maintained* emergency lighting system is one in which all emergency lamps are in operation
  (a) only when the normal lighting fails
  (b) at all material times
  (c) when a building is not occupied
  (d) and under control of the caretaker

210 The ohm is the unit of the following terms *except*
  (a) impedance
  (b) inductance
  (c) resistance
  (d) reactance

211 Which of the following is not an accessory?
  (a) luminaire
  (b) lampholder
  (c) adaptor
  (d) ceiling rose

212 An instrument used for measuring the effectiveness of earthing is called a
  (a) Murray loop tester
  (b) miltimeter
  (c) loop impedance tester
  (d) Wheatstone bridge

213 With all the switches closed in Figure 174
  (a) both lamps will not light
  (b) both lamps will light
  (c) only lamp L1 will light
  (d) only lamp L2 will light

214 In Figure 174 if all switches are closed and the link XY removed the lamps will be wired in
  (a) series
  (b) parallel
  (c) star formation
  (d) combination

215 The coulomb is the unit of
  (a) magnetic flux
  (b) magnetomotive force
  (c) electric charge
  (d) capacitance

216 The BS 3939 location symbol shown in Figure 175 is a
  (a) sub-main switch
  (b) main control point
  (c) starter
  (d) change-over switch

217 Which one of the following tools is used to mark lines on a metal object?
  (a) scriber
  (b) screwdriver
  (c) compass
  (d) hacksaw blade

218 A wattmeter is used for measuring power in
  (a) a.c. circuits only
  (b) d.c. circuits only
  (c) a.c. and d.c. circuits
  (d) non-electrical circuits

219 A low ohmic value resistor connected in parallel with a universal d.c. ammeter to extend its range is called a
  (a) multiplier
  (b) shunt
  (c) rheostat
  (d) variac

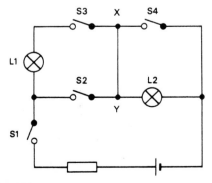

Figure 174

Figure 175

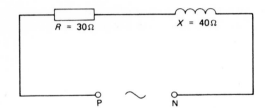

Figure 176

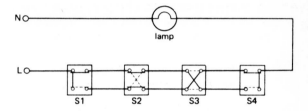

Figure 178

220 The total opposition to current flow in Figure 176 is called the
 (a) reactance
 (b) impedance
 (c) resistance
 (d) resistivity

221 The gas used in the design of a modern GLS lamp (Figure 177) is
 (a) oxygen
 (b) hydrogen
 (c) helium
 (d) nitrogen

222 The lamp circuit shown in Figure 178 provides
 (a) one-way and two-way switching control
 (b) two-way switching control
 (c) intermediate switching control
 (d) two-way and intermediate switching control

223 In Figure 179 to obtain resistance of 75 ohms the switch positions would be
 (a) S1 off, S2 on, S3 on
 (b) S1 on, S2 on, S3 off
 (c) S1 off, S2 on, S3 off
 (d) S1 on, S2 off, S3 on

224 The positive plate of a lead-acid cell on discharging changes to
 (a) lead peroxide
 (b) lead sulphate
 (c) spongy lead
 (d) water

225 Which combination of 2 V cells in Figure 180 provides the highest voltage output?
 (a) (b) (c) (d)

226 A PEN conductor is used throughout the distribution in a
 (a) IT system
 (b) TN-S system
 (c) TN-C system
 (d) TT system

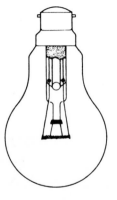

Figure 177

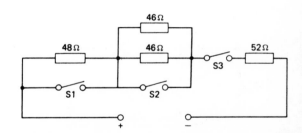

Figure 179

Revision exercise one

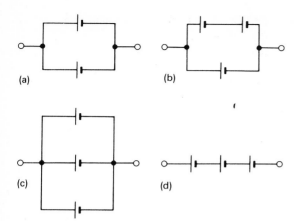

Figure 180

227 The current which will flow through a 1 ohm resistor connected to a primary cell of 1.4 volts having an internal resistance of 0.4 ohms is
(a) 2.0 A
(b) 1.4 A
(c) 1.0 A
(d) 0.4 A

228 The energy possessed by a body due to its motion is called
(a) electrical energy
(b) thermal energy
(c) kinetic energy
(d) potential energy

229 In Figure 181 the instruments connected to the circuit may be used to determine
(a) energy
(b) efficiency
(c) power factor
(d) frequency

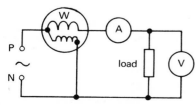

Figure 181

230 Where a residual current device provides protection against electric shock its sensitivity should be
(a) 13.0 A
(b) 3.0 A
(c) 0.3 A
(d) 0.03 A

231 What percentage of 110 V is 2.75 V?
(a) 10.0%
(b) 6.0%
(c) 2.5%
(d) 1.5%

232 A circuit is protected by a fuse rated at 5 A. If the fusing factor of the fuse is 1.2, the minimum fusing current will be
(a) 10.0 A
(b) 6.0 A
(c) 5.0 A
(d) 1.2 A

233 A transformer has a primary to secondary turns ratio of 25:1. If the primary voltage is 250 V, then the secondary voltage is
(a) 6.25 kV
(b) 250 V
(c) 25 V
(d) 10 V

234 Ignoring temperature change of the element, when a 3 kW, 240 V immersion heater is supplied at 120 V its power is reduced to
(a) 1500 W (c) 750 W
(b) 1000 W (d) 250 W

235 A resistor with a nominal value of 4000 ohms and tolerance of ± 5% would have a maximum value of
(a) 4200 ohms
(b) 4085 ohms
(c) 3800 ohms
(d) 3596 ohms

236 The efficacy of a 125 W fluorescent luminaire having control gear losses of 13 W with 8800 tube lighting design lumens is approximately
(a) 79 lm/watt
(b) 64 lm/watt
(c) 53 lm/watt
(d) 42 lm/watt

237 In a three-phase star-connected circuit, the ratio $\dfrac{\text{line volts}}{\sqrt{3}}$ gives
(a) the voltage to neutral or phase voltage
(b) the voltage across two lines of the supply
(c) the permissible volt drop of the system
(d) neutral/earth voltage

238 A force of 200 N lifts a machine 20 m from the ground: the work done is
(a) 40 kJ
(b) 10 kJ
(c) 4 kJ
(d) 1 kJ

239 Which of the following 13-A radial circuits in Figure 182 complies with the IEE Wiring Regulations for a domestic dwelling of floor area not exceeding 50 m² assuming the wiring system is MIMS cable?
(a) (b) (c) (d)

240 All phase colours of a three-core flexible cable should be identified with the colour
(a) brown
(b) orange
(c) red
(d) purple

241 The fusing current of a rewirable fuse is approximately
(a) twice its current rating
(b) three times its current rating
(c) four times its current rating
(d) five times its current rating

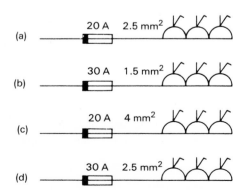

Figure 182

242 Distance X in Figure 183 is found to be
(a) 9 mm
(b) 8 mm
(c) 4 mm
(d) 3 mm

243 A ceiling rose should not be installed in any circuit operating at a voltage exceeding
(a) 500 V
(b) 415 V
(c) 320 V
(d) 250 V

244 What is the approximate volt drop on a 415 V consumer's premises if a limit of 2.5% is not to be exceeded?
(a) 14 V
(b) 10 V
(c) 6 V
(d) 2 V

245 Which of the following conduit wiring systems should not be installed in extremely cold environments?
(a) flexible metal conduit
(b) aluminium conduit
(c) plastic conduit
(d) steel conduit

246 The resistance of brass increases when the
(a) humidity increases
(b) temperature decreases
(c) temperature increases
(d) temperature remains constant

247 In Figure 184, the voltmeter would read
(a) 10 V
(b) 5 V
(c) between 0 V and 5 V
(d) 0 V

248 When two closely parallel conductors pass current in opposite directions, the effect will be to
(a) repel each other
(b) attract each other
(c) rotate about each other
(d) short-circuit each other

Revision exercise one

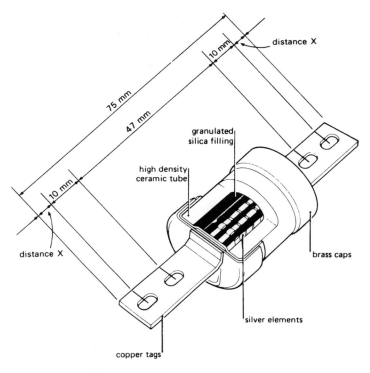

Figure 183

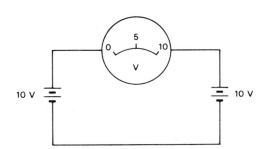

Figure 184

249 When lifting a heavy object from the ground, you should (1) bend your knees, and (2) lift the object by straightening your back
    (a)   both statements are true
    (b)   both statements are false
    (c)   only statement (1) is true
    (d)   only statement (2) is true

250 The volume of a copper rod 1 m long and 2 mm in diameter is
    (a)   3142 mm^3
    (b)   314.2 mm^3
    (c)   31.42 mm^3
    (d)   3.142 mm^3

251 An a.c. supply voltage of 240 V r.m.s. means that it
    (a)   has an average value of 120 V
    (b)   never rises above 240 V
    (c)   has an effective value of 240 V
    (d)   extinguishes itself twice every cycle

252 The force needed to stop and start things is called
    (a)   inertial force
    (b)   gravitational force
    (c)   cohesive force
    (d)   magnetic force

253 The rate at which work is done is called
    (a)   energy
    (b)   power
    (c)   force
    (d)   acceleration

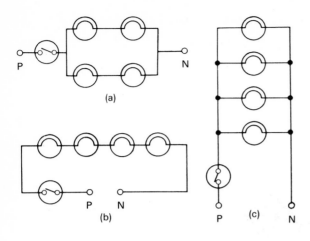

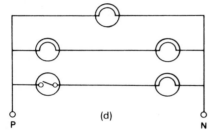

Figure 185

257 One advantage of using a three-phase, four-wire system over a three-phase, three-wire system is that
  (a) the system creates a star point to protect against lightning strokes
  (b) the system satisfies TNS requirements
  (c) there is no need to provide an earth protective conductor throughout the system
  (d) unbalanced currents created in the system can return through the neutral

258 The approximate cross-sectional area of a bare conductor 10 mm in diameter is
  (a) 92 mm^2
  (b) 85 mm^2
  (c) 79 mm^2
  (d) 63 mm^2

254 Which of the circuits shown in Figure 185 is normally used for a domestic lighting circuit?
  (a) (b) (c) (d)

255 The type of lamp shown in Figure 186 is called
  (a) an incandescent lamp
  (b) a fluorescent lamp
  (c) a halogen lamp
  (d) a sodium lamp

256 The exposed non-conducting metalwork of an electrical installation should be earthed in order to
  (a) avoid the flow of earth-leakage currents
  (b) prevent short-circuit faults
  (c) protect users against electric shock
  (d) reduce the risk of corrosion

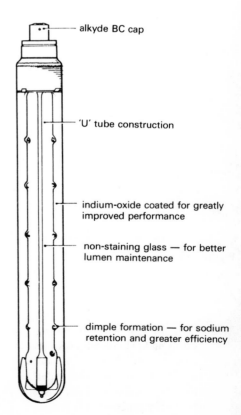

Figure 186

259 The inner radius of a 20 mm conduit bend should not be less than
(a) 50 mm
(b) 40 mm
(c) 30 mm
(d) 20 mm

260 In a three-phase, three-wire system, an open circuit fault in one phase of a connected load causes current in the other two phases to
(a) increase
(b) decrease
(c) remain constant
(d) fall to zero

261 The permitted volt drop allowed on consumers wiring supplied at 110 V is
(a) 6.00 V
(b) 5.25 V
(c) 2.75 V
(d) 1.33 V

262 In Figure 187, for each switch and light to operate independently the wiring circuit needs modifying so that the connection on terminal 3 is moved to terminal 2 and an additional wire links the terminals 3 and
(a) 1
(b) 5
(c) 6
(d) 4

263 The immediate first-aid treatment for a cut hand that is bleeding profusely is to
(a) wash the wound with warm water
(b) apply a tourniquet above the wound
(c) apply direct pressure on the wound
(d) keep the hand above the head

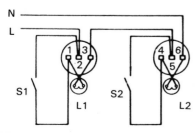

Figure 187

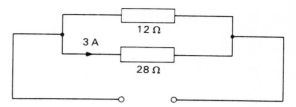

Figure 188

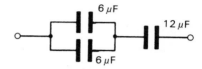

Figure 189

264 In Figure 188, the power dissipated in the 12 ohm resistor is
(a) 604 W
(b) 588 W
(c) 433 W
(d) 309 W

265 What is the correct colour identification for a three-wire d.c. supply?

	positive	middle	negative
(a)	red	black	blue
(b)	white	orange	black
(c)	red	white	blue
(d)	red	blue	black

266 The overall value of capacitance in Figure 189 is
(a) 15 µF
(b) 12 µF
(c) 6 µF
(d) 3 µF

267 If the capacitors shown in Figure 189 were connected to a supply of 500 V the charge conveyed in the circuit would be
(a) 3 MC
(b) 3 kC
(c) 3 mC
(d) 3 µC

268 In Figure 190 which shape has the greatest area?
   (a) (b) (c) (d)

269 The function of a double-pole linked switch is to
   (a) isolate or close both poles of the supply simultaneously
   (a) join one pole of the supply to the other in one movement
   (c) allow the switch to be used on both a.c. and d.c. supply
   (d) avoid short-circuiting both poles

270 The impedance of an a.c. circuit supplied at 275 kV and taking a load current of 0.5 kA is
   (a) 1375 Ω
   (b) 1155 Ω
   (c) 790 Ω
   (d) 550 Ω

271 The function of a muting switch on a zone indicator panel is to
   (a) silence the sound of a buzzer device
   (b) automatically call emergency services
   (c) inform workers of meal breaks
   (d) relay signals to other indicator panels

272 Unity power factor is when
   (a) power equals voltamperes
   (b) cos ϕ equals zero
   (c) reactance equals resistance
   (d) reactive power equals apparent power

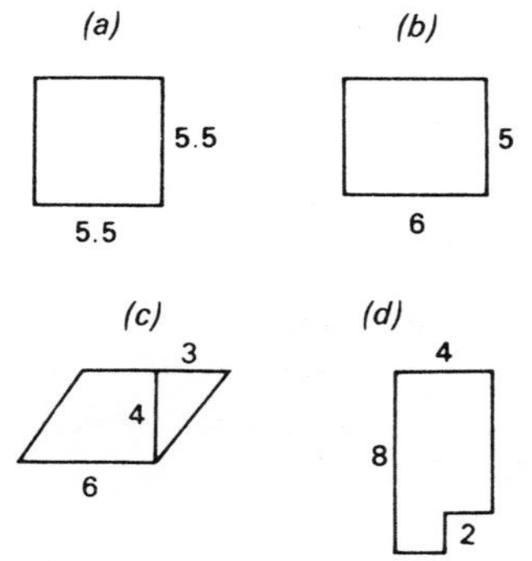

273 A battery of five 2-volt cells has a terminal p.d. of 9 V when supplying a current of 8 A. What is the internal resistance of each cell of the battery?
   (a) 125 mΩ
   (b) 75 mΩ
   (c) 40 mΩ
   (d) 25 mΩ

274 If the primary voltage of a single-phase transformer was 0.2 kV and its transformation ratio was 16:1, what would be its secondary voltage?
   (a) 3.2 kV
   (b) 1.6 kV
   (c) 0.025 kV
   (d) 0.0125 kV

275 In Figure 191 an earth fault occurs at point X. This will result in
   (a) any bell ringing whichever push is pressed
   (b) the fuse blowing before any push is pressed
   (c) all three bells ringing before any push is pressed
   (d) bell 2 and bell 3 becoming short-circuited together

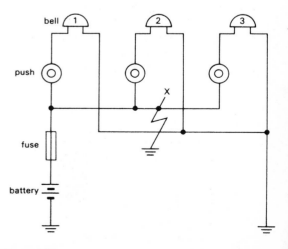

Figure 190

Figure 191

Revision exercise one

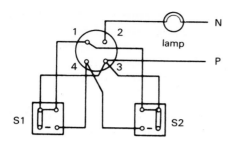

Figure 192

276 On a 50 Hz supply, a fluorescent tube extinguishes itself
(a) 200 times every second
(b) 100 times every second
(c) 50 times every second
(d) twice every second

277 In Figure 192, which terminal in the joint box is incorrectly connected?
(a) 4
(b) 3
(c) 2
(d) 1

278 Which of the following wiring systems is not particularly suitable for a farm installation?
(a) BE metal conduit containing PVC cables
(b) plastic conduit containing PVC cables
(c) cables sheathed with p.c.p. – not in contact with creosote
(d) cables sheathed with PVC – not in contact with creosote

279 To express kWh in basic SI units, 1 kWh is equal to
(a) 3.6 MJ
(b) 3.6 kJ
(c) 3.6 mJ
(d) 3.6 μJ

280 A 30-A domestic ring circuit may serve an unlimited number of 13-A socket-outlets provided the floor area does not exceed
(a) 200 m^2
(b) 100 m^2
(c) 50 m^2
(d) 10 m^2

281 Which of the following distribution voltages is standard in the UK?
(a) 450 V
(b) 440 V
(c) 415 V
(d) 400 V

282 How long will it take a 2.5 kW water heater to use 5 MJ of energy?
(a) 5000 s
(b) 2000 s
(c) 1250 s
(d) 500 s

283 In Figure 193 which switch position will produce the greatest heat output?
(a) (b) (c) (d)

284 In a d.c. generator, the purpose of the commutator is to
(a) rectify a.c. generated in the armature
(b) reduce eddy currents in the armature
(c) act as a speed mechanism for the armature
(d) reduce armature reaction

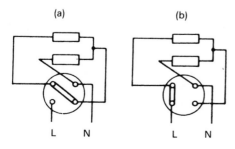
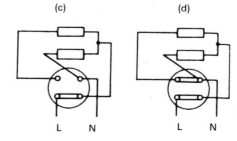

Figure 193

285 Fluxes are used in electrical soldering work to
 (a) protect a surface from oxidation
 (b) chemically wash a surface clean
 (c) protect the surface from corrosion
 (d) prevent the solder from running

286 Which BS 3939 symbol in Figure 194 identifies Figure 195?

287 In a factory supplied at 415 V, three-phase four-wire, the phase voltage would be
 (a) 415 V
 (b) 240 V
 (c) 110 V
 (d) 64 V

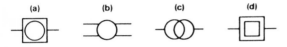

Figure 194

288 The object of making a visual inspection test on a portable appliance lead is to
 (a) check for cuts and abrasions
 (b) check for correct length
 (c) check its current rating
 (d) check its polarity

289 A residual current device should only be used where the product of its operating current and earth-loop impedance *does not exceed*
 (a) 50 V
 (b) 40 V
 (c) 30 V
 (d) 25 V

290 Which of the following is the heaviest for the same volume?
 (a) iron
 (b) brass
 (c) copper
 (d) lead

291 Figure 196 shows a 30-A ring circuit in PVC conduit. How many single-core cables will be run in the section marked XY?
 (a) 9
 (b) 6
 (c) 4
 (d) 3

292 In Figure 196 the spur socket-outlet to the right of the distribution board would be wired to the
 (a) other circuit spur
 (b) first socket-outlet on the ring
 (c) last socket-outlet on the ring
 (d) origin of circuit

293 Where an earth-leakage protective device protects an electrode water heater, the earthing conductor (being mechanically protected) must not be smaller than
 (a) 4.0 mm^2
 (b) 2.5 mm^2
 (c) 1.5 mm^2
 (d) 1.0 mm^2

294 Soft solder used for electrical joints is an alloy of
 (a) tin and zinc
 (b) zinc and copper
 (c) tin and lead
 (d) lead and zinc

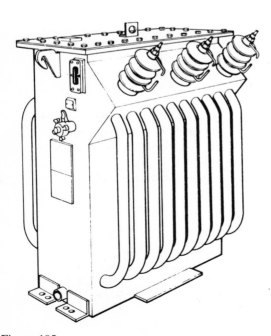

Figure 195

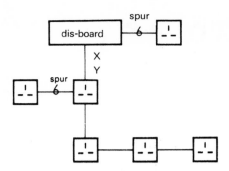

Figure 196

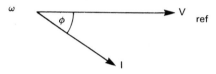

Figure 197

295 The formula *length of run (m) × load current (I) × millivolt/A/m (mV)* is used in cable selection procedures to find
(a) permissible volt drop
(b) actual volt drop
(c) assumed volt drop
(d) rated volt drop

296 The purpose of PVC covering on MIMS cable is to protect the metal sheath against
(a) sunlight (c) corrosion
(b) lightning (d) vermin

297 To stop a permanent magnet losing its magnetism, its poles are
(a) left uncovered
(b) smothered in grease
(c) bitumenized
(d) shorted by soft iron

298 Which of the following components when connected to an a.c. supply would produce a phasor diagram as shown in Figure 197?
(a) fire-bar element
(b) filament lamp
(c) choking coil
(d) capacitor

299 Which of the following is excluded from the scope of the IEE Wiring Regulations?
(a) Hospital installations
(b) Safety sources – emergency lighting
(c) Multi-storey lateral mains
(d) Garages – safe area circuits

300 In a domestic dwelling, a portable heating appliance can be used in the following areas *except*
(a) landing
(b) kitchen
(c) bathroom
(d) attached garage

Revision exercise two

301 In Figure 198 the thyristor will conduct over the periods marked
 (a) 1–2 and 5–6
 (b) 1–3 and 5–7
 (c) 2–3 and 6–7
 (d) 2–4 and 6–8

302 Which are the BS 4999 markings for the connections of a three-phase induction motor internally delta connected?
 (a) L1 L2 L3
 (b) U V W
 (c) A1 A2 A3
 (d) X Y Z

303 In a p-type semiconductor material the holes are called
 (a) majority current carriers
 (b) minority current carriers
 (c) covalent bonds
 (d) positive ions

304 The recommended standard service illuminance value for hand engraving is
 (a) 4000 lux
 (b) 3000 lux
 (c) 2000 lux
 (d) 1000 lux

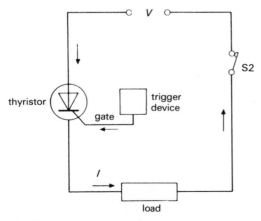
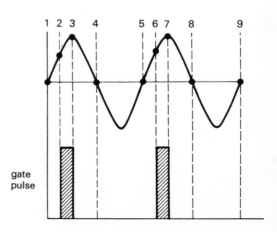

Figure 198

Revision exercise two

305 A waveformal CNE cable has its conductors insulated with
(a) cross-linked polyethylene
(b) tough rubber sheath
(c) mineral insulation
(d) paper insulation

306 To reverse the direction of a single-phase capacitor-start induction motor one must interchange the
(a) centrifugal switch connections
(b) capacitor connections
(c) supply leads
(d) starting winding leads

307 When starting a three-phase induction motor using a star-delta starter, the line current is reduced by approximately
(a) ¾
(b) ⅔
(c) ⅓
(d) ¼

308 What is the code for a high-pressure sodium vapour lamp?
(a) SOX
(b) SON
(c) SHP
(d) SVL

309 In terms of days, the critical path of the project shown in Figure 199 is the line representing
(a) 25.5 days
(b) 18.0 days
(c) 17.0 days
(d) 15.0 days

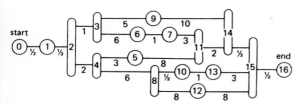

Figure 199

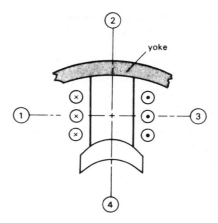

Figure 200

310 Which numbered circle identifies the south-seeking end of the salient-pole electromagnet shown in Figure 200?
(a) 1
(b) 2
(c) 3
(d) 4

311 The previous and present energy meter readings in a domestic dwelling are 44157 and 45409 respectively. If the standing charge per quarter is £5.07, the consumer's quarterly bill at 4.18p per unit, is
(a) £67.66 (c) £43.25
(b) £57.40 (d) £31.87

312 The stator core magnetic field in a three-phase induction motor travels at
(a) rotating speed
(b) asynchronous speed
(c) synchronous speed
(d) slip speed

313 When a.c. induction motors are lightly loaded their power factors are
(a) lowered (c) unchanged
(b) increased (d) non-existent

314 An alternating e.m.f. has a frequency of 50 Hz. What is the time over one cycle?
(a) 5.00 s (c) 0.25 s
(b) 2.50 s (d) 0.02 s

315 The no-volt release coil inside the face-plate starter of a d.c. shunt motor is sometimes fed from the
(a) internal armature resistance
(b) internal overload trip
(c) external field supply
(d) external mains supply

316 Which of the diagrams in Figure 201 represents an isolating transformer?
(a) (b) (c) (d)

317 Growth factor is the ratio of
(a) $\dfrac{\text{estimated future load}}{\text{maximum load}}$
(b) $\dfrac{\text{average load}}{\text{maximum load}}$
(c) $\dfrac{\text{distance moved by effort}}{\text{distance moved by load}}$
(d) $\dfrac{\text{mechanical output}}{\text{electrical input}}$

318 A device used to prevent a temperature rise in a semiconductor rectifier is called a
(a) thermocouple
(b) heat sink
(c) rheostat
(d) surge divertor

319 The specific heat capacity of water is
(a) 5800 J/kg °C
(b) 4200 J/kg °C
(c) 3600 J/kg °C
(d) 2000 J/kg °C

320 The ampere-hour efficiency of a secondary cell is 80%. If the Ah on discharge is 50, the Ah on charge is
(a) 85.0
(b) 74.5
(c) 62.5
(d) 57.0

321 Forward bias in a semiconductor device is when the potential barrier in the depletion layer is
(a) increased
(b) reduced
(c) constant
(d) saturated

322 What is the maximum value reached by an a.c. supply of 415 V?
(a) 931 V
(b) 829 V
(c) 654 V
(d) 587 V

323 A conductor 15 mm in length, lies at right angles to a magnetic field of strength 15 T. What is the force exerted on it when carrying a current of 500 mA?
(a) 112.5 MN
(b) 112.5 kN
(c) 112.5 mN
(d) 112.5 μN

324 What is the approximate volt drop in a three-core cable having a resistance per core of 11 mΩ and carrying a current of 500 A to a balanced three-phase a.c. load?
(a) 16.5 V
(b) 11.0 V
(c) 10.4 V
(d) 9.5 V

325 Which of the following wiring systems is *not* recommended for a petrol filling station?
(a) flexible metal conduit incorporating PVC cables
(b) solid-drawn, heavy-gauge metal conduit incorporating PVC cables
(c) MIMS cables with overall PVC sheath
(d) armoured PVC cables

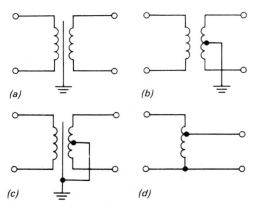

Figure 201

Revision exercise two

326 The maximum horizontal spacing between space-bar saddles supporting 20 mm rigid metal conduit is
(a) 3.00 m
(b) 2.85 m
(c) 2.50 m
(d) 1.75 m

327 The purpose of a circuit chart inside a distribution board is to
(a) provide information on the route taken by the load
(b) provide details about the class of excess-current protection used
(c) indicate to the user the size of fuse element fitted
(d) indicate the circuit connected to each fuse-way

328 Figure 202 shows the four elements of a bridge-connected rectifier. To which terminals will the a.c. supply be connected?
(a) 1 and 3
(b) 2 and 4
(c) 3 and 5
(d) 4 and 1

329 In Figure 202 where would you connect the d.c. negative load terminal?
(a) 1
(b) 2
(c) 3
(d) 4

330 In Figure 202 to provide full-wave d.c., the terminals that require joining together are
(a) 1 and 2
(b) 1 and 3
(c) 1 and 4
(d) 1 and 5

331 A 1000 m length of cable has an insulation resistance of 50 MΩ. What is its insulation resistance for 100 m?
(a) 500 MΩ
(b) 100 MΩ
(c) 50 MΩ
(d) 5 MΩ

332 It is recommended that oil filled apparatus be installed in a building of fire-resistant construction, with external ventilation, if their oil capacity exceeds.
(a) 35 l
(b) 30 l
(c) 25 l
(d) 20 l

333 If a three-phase motor is run up to speed and, for example, a fuse blows, the motor will
(a) slow down and stop because it has no travelling magnetic force
(b) continue to run but overload the good phases
(c) stop immediately because of single-phasing
(d) run faster than normal since it operates with fewer pole pairs

334 The effect of the fuse blowing (see question 333) or a phase winding burning out is called
(a) armature reaction
(b) split-phasing
(c) single-phasing
(d) hunting

335 If a wound-rotor induction motor had its slip-rings open-circuited at the point of switching on the supply, it would
(a) refuse to start altogether
(b) start and then stop immediately
(c) start and run continuously
(d) start and stop intermittently

336 Figure 203 shows the test for
(a) verification of polarity
(b) element continuity
(c) earth impedance
(d) insulation resistance

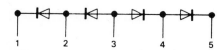

Figure 202

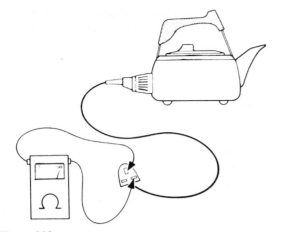

Figure 203

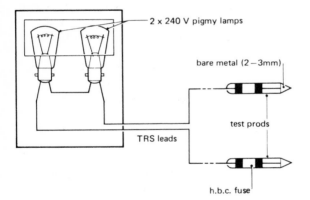

Figure 204

337 The size of metal conduit for carrying ten solid pvc 1.5 mm² cables over a length of 5 m incorporating two bends is
(a) 32 mm
(b) 25 mm
(c) 20 mm
(d) 16 mm

338 The Health and Safety at Work etc. Act (1) covers all people at work, and (2) excludes domestic servants in private households
(a) both statements (1) and (2) are correct
(b) both statements (1) and (2) are incorrect
(c) only statement (1) is correct
(d) only statement (2) is correct

339 The object of wiring two 240 V pigmy lamps in series in the test lamp set shown in Figure 204 is to allow it to be used
(a) on a.c. and d.c. supplies up to 500 V
(b) to give different lamp brightness for different voltages
(c) safely in the event of an open-circuit fault
(d) safely in the event of an internal short-circuit fault

340 Switches used to control discharge-lighting circuits should be rated for
(a) four times the steady circuit current
(b) three times the steady circuit current
(c) twice the steady circuit current
(d) one and a half times the steady circuit current

341 Approximate unity power factor condition is achieved from the following apparatus *except*
(a) incandescent lamps
(b) fluorescent lead-lag twin-lamps
(c) radiant heaters
(d) resistance-start induction motors

342 Five motors take full load currents of 25 A, 20 A, 15 A, 10 A, and 10 A respectively. What is the total estimated current after applying the following diversity factors? (1) 100% full load of the largest motor; (2) 80% full load of the second largest motor; and (3) 60% full load of the remaining motors
(a) 80 A
(b) 75 A
(c) 62 A
(d) 48 A

343 A linear double-ended tungsten-halogen lamp will have its life reduced if
(a) it is fitted with an elliptical reflector
(b) the supply voltage falls below 2.5%
(c) not operated in the vertical position
(d) not operated in the horizontal position

344 For remote control of a direct-on-line contactor starter
(a) start buttons are wired in series and stop buttons are wired in series
(b) start buttons are wired in parallel and stop buttons are wired in parallel

(c) start buttons are wired in series and stop buttons are wired in parallel
(d) start buttons are wired in parallel and stop buttons are wired in series

345 The number of lighting outlet points that can be connected to a final circuit controlled by a 5-A fuse (assuming a 240 V supply) is
(a) 12
(b) 10
(c) 9
(d) 8

346 Every re-inspection of an electrical installation should be carried out by a professionally qualified and/or competent person belonging to or acting for one of the following except
(a) Electrical Contractors' Association
(b) National Inspection Council for Electrical Installation Contracting
(c) Royal Institute of British Architects
(d) Institution of Electrical Engineers

347 The maximum length of span and minimum height above ground for pvc cables in heavy gauge conduit, installed inaccessible to vehicular traffic is
(a) 8 m
(b) 5 m
(c) 3 m
(d) 2 m

348 The magnitude of fault current that is likely to flow under extreme conditions of negligible fault impedance is called
(a) nominal earth current
(b) overload current
(c) prospective short-circuit current
(d) earth-loop path current

349 A thermal cut-off device is used to
(a) control temperature variations
(b) protect against overheating
(c) measure air temperatures
(d) safeguard against frost

350 In Figure 205, the function of the copper shading ring is to
(a) delay the flux in the main pole
(b) improve commutation
(c) stop sparking on the rotor
(d) reduce eddy currents

351 In a 415 V contactor, the coil hums but the contactor does not close. The cause of the fault could be that
(a) the start push-button is sticking
(b) the supply fuse is blown
(c) an earth fault is on one phase
(d) there is dirt on the magnet faces

352 A d.c. supply to a fluorescent lamp commonly employs
(a) an autotransformer
(b) an inverter ballast
(c) an infra-red detector
(d) a mercury-arc rectifier

353 Which of the following lamps has the lowest efficacy value?
(a) 500 W tungsten lamp (GLS)
(b) 400 W high-pressure sodium lamp (SON)
(c) 250 W high-pressure mercury lamp (MBF/U)
(d) 125 W fluorescent tube (MCF)

354 An incandescent lamp has a luminous intensity of 150 cd and gives an illuminance of 1.5 lx on the working plane directly below. What is the illuminance on the working plane if the lamp's distance is halved?
(a) 30 lx
(b) 10 lx
(c) 6 lx
(d) 3 lx

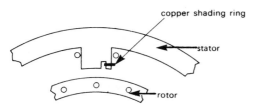

Figure 205

355　A three-phase cage induction motor suffers from overheating. The following are likely causes *except*
(a)　high ambient temperature
(b)　dust in the ventilation openings
(c)　misalignment
(d)　supply regeneration

356　The visual inspection of an electrical installation is made to confirm the condition of
(a)　conductor continuity
(b)　electrical equipment
(c)　insulation resistance
(d)　earth impedance

357　Which of the following circuits in Figure 206 shows the correct method of connecting an ammeter and voltmeter through instrument transformers?
(a)　(b)　(c)　(d)

358　One advantage of a TN-C-S system is that it enables the supply authority to provide
(a)　a safe and reliable method of earthing consumers' premises
(b)　consumers with a star-point connection to facilitate earthing
(c)　earthing to all parts of a consumer's electrical wiring via electrodes
(d)　an earthing terminal to all consumers' premises protected by earth-leakage circuit breakers

359　One disadvantage of a high-voltage autotransformer is that if an open-circuit fault develops in the common part of its winding, a
(a)　dangerous voltage will appear on the secondary side
(b)　dangerous voltage will appear on the primary side
(c)　voltage will not appear on the load side
(d)　current will not flow in the load circuit

360　What is the back e.m.f. of a d.c. motor having an armature resistance of 0.15 Ω and taking a current of 20 A from a 220 V supply?
(a)　240 V
(b)　223 V
(c)　217 V
(d)　210 V

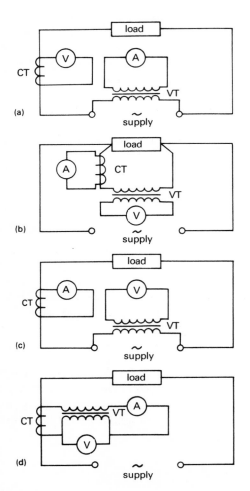

Figure 206

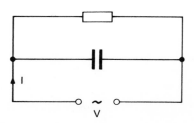

Figure 207

361 Which of the following phasor diagrams represents Figure 207?

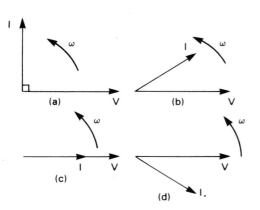

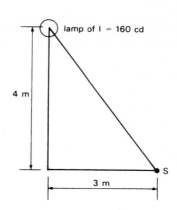

Figure 208

362 The power input of an electric pump which lifts 20 litres of water through a height of 5 metres in 12.26 seconds having an efficiency of 80% is, approximately
(a) 500 W
(b) 250 W
(c) 135 W
(d) 100 W

363 The illuminance at point S in Figure 208 is
(a) 25.60 lx
(b) 17.30 lx
(c) 6.40 lx
(d) 5.12 lx

364 A single phase, double-wound transformer having primary and secondary turns of 250 and 50 respectively, takes a primary current of 20 A. What current is taken by the secondary circuit?
(a) 250 A
(b) 100 A
(c) 50 A
(d) 4 A

365 In Figure 207, if the resistor passed was 6 A and the capacitor passed 8 A, what would be the current taken from the supply?
(a) 14 A
(b) 10 A
(c) 8 A
(d) 6 A

366 In Figure 207, if the resistor was replaced by an inductor of negligible resistance and took a current of 8 A, the current taken from the supply would be
(a) 16 A
(b) 11 A
(c) 8 A
(d) 0 A

367 Which of the graphs in Figure 209 shows variation of capacitive reactance with frequency?
(a) 1
(b) 2
(c) 3
(d) 4

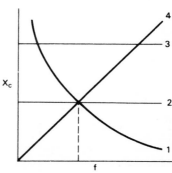

Figure 209

368 If a 10 kVA transformer supplies a load of 5 kW at a power factor of 0.5, then it will be delivering
   (a) full load
   (b) half full load
   (c) twice full load
   (d) one quarter full load

369 A moving coil instrument gives full-scale deflection with a current of 15 mA and has a resistance of 5 Ω. What is the value of a shunt to allow the instrument to read up to 5 A?
   (a) 75 mΩ
   (b) 42 mΩ
   (c) 15 mΩ
   (d) 5 mΩ

370 A diagram showing the variation of luminous intensity with angle is called a
   (a) flow diagram
   (b) Gannt chart
   (c) Venn diagram
   (d) polar curve

371 The BS 3939 graphical symbol in Figure 210 shows
   (a) two telephone points
   (b) an emergency (safety) lighting point
   (c) a make and break contact point
   (d) a normally closed interlock point

Figure 210

372 A new cable on the market, not covered by existing British Standards, may be given a certificate after the approval of the IEE Wiring Regulations Committee. Which of the following might this be?
   (a) a CENELEC certificate
   (b) a BASEC certificate
   (c) a BSI certificate
   (d) an MTIRA certificate

373 A fused plug to the design of BS 1363 must not be fitted with a fuse in excess of
   (a) 30 A
   (b) 15 A
   (c) 13 A
   (d) 10 A

374 Which of the diagrams in Figure 211 represents the correct internal linking inside the terminal cover of a d.c. shunt motor?
   (a) (b) (c) (d)

375 Cables of category 1 circuits cannot be drawn into the same conduit as cables of category 2 circuits unless the latter are
   (a) insulated to the highest voltage present in category 1 circuits
   (b) insulated with mineral insulation throughout their length
   (c) not operating in ambient temperatures above 30°C
   (d) not operating on frequencies above 50 Hz

376 *The direction of an induced e.m.f. is always such that it tends to set up a current opposing the motion or the change of flux responsible for inducing that e.m.f.* This is known as
   (a) Ohm's law
   (b) Kirchhoff's law
   (c) Faraday's law
   (d) Lenz's law

377 The ratio of illumination from a dirty installation to that from the same installation when clean is called the
   (a) maintenance factor
   (b) diversity factor
   (c) correcting factor
   (d) rating factor

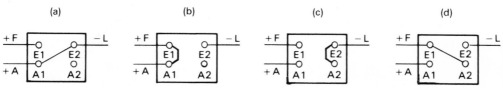

Figure 211

Revision exercise two 171

378 The power dissipated by a machine tool part is 2 kW. If the part revolves at a speed of 79.57 rev/s, its torque is
(a) 25 Nm
(b) 16 Nm
(c) 12 Nm
(d) 4 Nm

379 The maximum length of span for a PVC covered overhead line between buildings is
(a) 50 m
(b) 30 m
(c) 10 m
(d) 5 m

380 The letter designation classifying a PME system is
(a) TN–C
(b) TN–C–S
(c) TN–S
(d) TT

381 Before an ammeter is removed from the circuit of an instrument current transformer (c.t.), the secondary side of the c.t. requires shorting out. This is to avoid
(a) a dangerously high voltage occurring across the transformer terminals
(b) damage to the ammeter from stray magnetic fields
(c) wrongful connection of the transformer when the ammeter is replaced
(d) damage to the transformer due to travelling voltage surges

382 A ring main system is a
(a) primary supply distribution system
(b) category 3 wiring system
(c) high-voltage alarm system
(d) system of wiring 13-A socket-outlets

383 Until magnetic saturation is reached, torque of a d.c. series motor is proportional to
(a) I^2
(b) R^2
(c) V^2
(d) E^2

where I is armature current
R is armature resistance
V is supply voltage
E is back e.m.f.

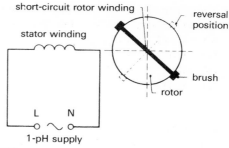

Figure 212

384 The name given to the motor shown in Figure 212, whose commutator brushes are shorted out is
(a) universal motor
(b) repulsion motor
(c) split-phase motor
(d) shaded-pole motor

385 In Figure 209 on page 169, if line 1 represents X_C and line 4 represents X_L, the cross-over point is called
(a) resonant frequency
(b) minimum frequency
(c) average frequency
(d) zero frequency

386 The output power of an a.c. motor *cannot* develop if its
(a) torque is unity
(b) slip is zero
(c) speed is above synchronous speed
(d) rotor is not moving

387 In Figure 213, the current taken by the circuit is
(a) 50 A
(b) 25 A
(c) 10 A
(d) 5 A

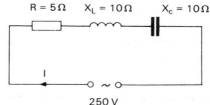

Figure 213

388 The per unit slip of an a.c. induction motor having a synchronous speed of 25 rev/s and rotor speed of 23.75 rev/s, is
(a) 0.05
(b) 0.03
(c) 0.02
(d) 0.00

389 A wattmeter instrument can be used on
(a) a.c. supplies but not d.c. supplies
(b) d.c. supplies but not a.c. supplies
(c) both a.c. and d.c. supplies
(d) a.c. supplies not exceeding 415 V

390 The resistance of a copper conductor of c.s.a. 2.5 mm^2, length 100 m and resistivity 17.8 $\mu\Omega$mm, is
(a) 712 MΩ
(b) 712 kΩ
(c) 712 Ω
(d) 712 mΩ

391 The sectional insulation resistance values for a completed electrical installation are 40 MΩ, 30 MΩ, 2 MΩ, and 1 MΩ respectively. The combined test value would be
(a) 73.00 MΩ
(b) 40.00 MΩ
(c) 1.00 MΩ
(d) 0.64 MΩ

392 One advantage of an m.c.b. over other forms of excess-current protection is that
(a) its on-off position can be easily identified
(b) its operation is silent over a wide range of fault levels
(c) it operates on both earth fault and short-circuit fault
(d) it is easily installed and reset quickly

393 The specific gravity of an accumulator is directly measured using a
(a) voltmeter
(b) calorimeter
(c) hydrometer
(d) density bottle

394 The secondary winding of a current instrument transformer must not be opened while the primary winding is carrying current, because a
(a) short-circuit will result in the primary winding
(b) high e.m.f. may be induced in the secondary winding
(c) high-voltage surge will propagate in the primary circuit
(d) break in the circuit will cause magnetic saturation of the c.t. core

395 In a fluorescent tube switch-start circuit, a short-circuit occurs in the starter causing the tube to
(a) glow brightly at one end
(b) glow brightly at both ends
(c) flicker on and off
(d) go out

396 The equation

$$\frac{\text{total lamp lumens} \times \text{utilization factor}}{\text{area}}$$

gives the
(a) glare index
(b) room index
(c) efficacy
(d) lux level

397 The two protective devices shown in Figure 214 are wired in series with each other. For a prospective current indicated by XY
(a) the 5-A fuse will operate first
(b) both devices will operate together
(c) the 5-A m.c.b. will operate first
(d) both devices will not operate

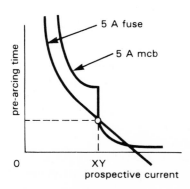

Figure 214

398 A material commonly used to support flat busbars in a busbar chamber is
(a) hard rubber
(b) Bakelite
(c) porcelain
(d) asbestos

399 With reference to Figure 215, the speed-torque characteristic is typical for a
(a) universal motor
(b) d.c. shunt motor
(c) d.c. series motor
(d) a.c. repulsion motor

400 Under the 'Conditions of Supply', earthing of a consumer's installation is the responsibility of the

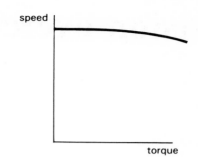

Figure 215

(a) area electricity board
(b) Central Electricity Generating Board
(c) local authority
(d) owner of the premises

Revision exercise three

IEE Wiring Regulations Assessment

401 Which one of the following is NOT a statutory document?
(a) The Health and Safety Act
(b) The Electricity Supply Regulations
(c) The Electricity (Factory Act) Special Regulations
(d) The IEE Wiring Regulations

402 A TT system is one that relies on earth leakage current travelling back to the supply source via
(a) sheath of service cable
(b) earth electrode in soil
(c) aerial earth wire
(d) underground pipes

403 Which of the following is NOT a Category 1 Circuit?
(a) Emergency lighting circuit
(b) Domestic ring final circuit
(c) Industrial motor/starter circuit
(d) Caravan supply circuit

404 To satisfy the Supply Regulations to avoid excess leakage current to earth, a test on the electric wiring is made for
(a) insulation resistance
(b) conductor resistance
(c) conductor impedance
(d) earth fault loop impedance

405 The maximum permitted volt drop allowed on a domestic installation is
(a) 6.0%
(b) 5.0%
(c) 4.8%
(d) 2.5%

406 Class II equipment is designed
(a) flameproof
(b) double insulated
(c) instrinsically safe
(d) waterproof

407 Which of the following protective devices suffers from having a relatively high fusing factor?
(a) BS 88 Part 2 fuse
(b) BS 3036 fuse
(c) BS 1361 fuse
(d) BS 3871 Part 1 m.c.b.

408 A residual current device should only be used where the product of its operating current and earth loop impedance does NOT exceed
(a) 50 V
(b) 40 V
(c) 30 V
(d) 25 V

409 6.25 V is 2.5% of
(a) 415 V
(b) 250 V

(c) 240 V
(d) 230 V

410 The estimated external impedance (Z_E) for a TN-C-S system is
(a) 1.8 Ω
(b) 0.8 Ω
(c) 0.35 Ω
(d) 0.02 Ω

411 The time to interrupt a short circuit is given by the expression:
(a) $t = \dfrac{kS}{I^2}$
(b) $t = \dfrac{k^2S^2}{I^2}$
(c) $t = \dfrac{\sqrt{S}}{I}$
(d) $t = \dfrac{\sqrt{kS}}{I}$

412 Where protection against fire is required without a fire-resistant screen installed, the minimum clearance allowed above the heat source is
(a) 1.0 m
(b) 800 mm
(c) 500 mm
(d) 300 mm

413 Where a shower cubicle is located in a room other than a bathroom, *all* socket outlets (except shaver s.o.) should be installed more than
(a) 4.0 m away from the shower cubicle
(b) 3.0 m away from the shower cubicle
(c) 2.5 m away from the shower cubicle
(d) 1.0 m away from the shower cubicle

414 The 15th Edition of the IEE Regulations does NOT directly relate to
(a) caravan sites
(b) construction sites
(c) farm installations
(d) offshore installations

415 The current demand to be assumed for a lighting outlet is based on a minimum lampholder wattage of
(a) 500 W
(b) 250 W
(c) 100 W
(d) 60 W

416 What is the current demand for six 80 W fluorescent luminaires fed from a 240 V supply having a power factor of 0.5 lagging?
(a) 7.2 A
(b) 5.0 A
(c) 4.0 A
(d) 3.6 A

417 The recommended interval between inspections for a fire alarm installation is
(a) 5 years
(b) 1 year
(c) 6 months
(d) 3 months

418 The insulation resistance of a certain cable is 100 MΩ per km. For a 250 m length of the same cable its insulation resistance value will be
(a) 2500 MΩ
(b) 400 MΩ
(c) 250 MΩ
(d) 40 MΩ

419 When testing the insulation resistance of large installations, the outlets may be divided into groups containing not less than
(a) 50 outlets
(b) 40 outlets
(c) 35 outlets
(d) 20 outlets

420 A circuit is protected by a fuse rated 5 A. If the fusing factor of the fuse is 1.2, its minimum fusing current will be
(a) 10
(b) 6
(c) 5
(d) 1.2

421 An installation is earthed to
(a) protect against short circuit faults
(b) prevent exposed metal becoming live
(c) enable polarity tests to be made
(d) prevent a voltage appearing on the neutral conductor

422 An earthed concentric wiring system is one in which
(a) the internal conductor is negative to the external conductor

(b) the resistance of the external conductor is greater than that of the internal conductor
(c) the line and earth are both suitably fused
(d) the outer metallic sheath is used for the earth return

423 Which of the following earthing distribution systems is known as TN-C-S? (See Figure 216).

424 To obtain an accurate measurement of earth electrode resistance it is essential that the
(a) current distribution is concentric
(b) resistance areas do not overlap
(c) surface potential is zero
(d) current distribution is radial

425 The system shown below in Figure 217 is called a TT system. From point E, a connection is taken to
(a) a separate earth electrode
(b) the neutral of the supply
(c) the supply earth electrode
(d) the point marked C on the diagram

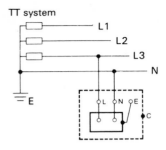

Figure 217

426 A fixed final circuit is protected by a BS 3036 fuse. If $I_b = 33$ A, what size fuse is suitable?
(a) 60 A
(b) 45 A
(c) 30 A
(d) 20 A

427 In Q426, if the ambient temperature correction factor was 0.94, what value of I_z needs to be chosen?
(a) 66 A
(b) 47 A
(c) 33 A
(d) 30 A

428 In Q426, what is the maximum earth fault loop impedance?
(a) 1.2 Ω
(b) 1.6 Ω
(c) 2.8 Ω
(d) 4.0 Ω

429 In Q426, if the final circuit was changed to feed a socket outlet, what would be the maximum earth fault loop impedance?
(a) 0.4 Ω
(b) 0.6 Ω
(c) 1.1 Ω
(d) 1.8 Ω

430 In Q429, what would be the maximum disconnection time allowed?
(a) 400 ms
(b) 500 ms
(c) 0.4 ms
(d) 0.5 ms

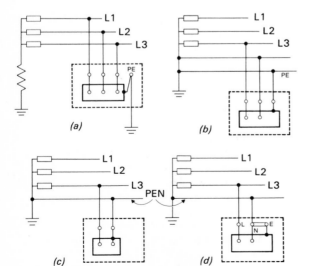

Figure 216

Revision exercise three

431 The ratio $\dfrac{\text{minimum actual load}}{\text{installed load}}$ is called
    (a) correction factor
    (b) rating factor
    (c) growth factor
    (d) diversity factor

432 The IEE Wiring Regulations tables are based on an ambient temperature of
    (a) 30°C
    (b) 35°C
    (c) 40°C
    (d) 50°C

433 In the Electricity (Factory Act) Special Regulations, a competent person may have a minimum age of
    (a) 25 years
    (b) 21 years
    (c) 18 years
    (d) 16 years

434 The red 'live' conductor of a single phase circuit is now called
    (a) potential conductor
    (b) positive conductor
    (c) phase conductor
    (d) line conductor

435 The term 'circuit protective conductor' now replaces the term
    (a) earthing lead
    (b) earth electrode conductor
    (c) bonding conductor
    (d) earth continuity conductor

436 A temporary installation on a construction site should be tested every
    (a) 5 years
    (b) 3 years
    (c) 6 months
    (d) 3 months

437 A 6 mm^2 PVC twin and earth armoured cable has a design current of 40 A and a 7.4 mV drop/A/m. What is the cable's actual voltage drop if its length of run is 30 m?
    (a) 8.88 V
    (b) 6.00 V
    (c) 2.22 V
    (d) 1.20 V

438 Using diversity allowance, what is the assessed design current of a 14 kW/240 V domestic cooker assuming the control panel incorporates a 13-A socket outlet?
    (a) 58.3 A
    (b) 53.3 A
    (c) 42.6 A
    (d) 29.5 A

439 Using Table 54F of the IEE Regulations, the minimum protected conductor size for a phase conductor of 25 mm^2 is
    (a) 50 mm^2
    (b) 25 mm^2
    (c) 16 mm^2
    (d) 10 mm^2

440 The minimum size of supplementary bonding conductor permitted in a bathroom where no mechanical protection is provided is
    (a) 4.0 mm^2
    (b) 2.5 mm^2
    (c) 1.5 mm^2
    (d) 1.0 mm^2

441 Unless otherwise stated by the local fire authority, a fireman's switch should not be higher than
    (a) 3.00 m
    (b) 2.75 m
    (c) 1.80 m
    (d) 1.24 m

442 When testing a portable drill for insulation resistance, the ohmic value should not be less than
    (a) 500 Ω
    (b) 0.5 Ω
    (c) 0.5 MΩ
    (d) 1.0 MΩ

443 Which of the following is not suitable for switching off for mechanical maintenance?
    (a) circuit breakers
    (b) thyristors
    (c) plugs and sockets
    (d) control switches

444 In a 240 V metal-clad fuseboard, the phase and neutral busbars are protected by insulation and by barriers with all live parts fully shrouded against direct contact. In these circumstances, the fuseboard may be
(a) opened without the use of a key or tool
(b) left unearthed
(c) used to accommodate SELF circuits
(d) used to accommodate Bell circuits

445 For household premises the floor area served by a single 30 A ring circuit using BS 1363 sockets may be
(a) up to 200 m^2
(b) up to 100 m^2
(c) up to 30 m^2
(d) unlimited

446 A certain cable has a design current of 4 A and a volt drop of 15 mV/A/m. If maximum voltage drop is allowed, the longest length of cable which can be used is
(a) 500 mm
(b) 340 mm
(c) 100 mm
(d) 25 mm

447 The cable factor for ten 10-mm^2 single-core PVC cables incorporating bends is
(a) 1460
(b) 1050
(c) 2690
(d) 3630

448 Which of the following is a statutory document?
(a) The Electricity Supply Regulations 1988
(b) BS 1013 Earthing 1965
(c) CP 1008 Maintenance of Electrical Switchgear
(d) Home Office Model Code for Storage of Petroleum Gas

449 The minimum size of PVC-insulated copper conductor to supply a radial circuit having four 13 A socket outlets in a room area of 35 m^2 is
(a) 1.0 mm^2
(b) 1.5 mm^2
(c) 2.5 mm^2
(d) 4.0 mm^2

450 Which of the following is not a method of protection against direct contact?
(a) Protection by insulation of live parts
(b) Protection by placing out of reach
(c) Protection by obstacles
(d) Protection by electrical separation

451 Where bonding is carried out on a gas meter, the connection should be made within
(a) 0.6 m
(b) 0.5 m
(c) 0.4 m
(d) 0.3 m

452 The phase conductor of a circuit is 50 mm^2. The minimum c.s.a. of the corresponding protective conductor is
(a) 50 mm^2
(b) 45 mm^2
(c) 30 mm^2
(d) 25 mm^2

453 A caravan inlet point should comply with
(a) BS 4052
(b) BS 4343
(c) BS 5050
(d) BS 6742

454 The electrical term 'spur' means
(a) branch cable
(b) 13-A outlet box
(c) ring main link
(d) junction box connection

455 Immediate action in the case of a person suffering electric shock is to
(a) seek medical assistance
(b) apply artificial respiration
(c) switch off supply
(d) shout for help

456 What is the current taken by a resistive load of 400 Ω and consuming 10 kW?
(a) 25 A
(b) 15 A
(c) 10 A
(d) 5 A

457 The maximum distance between supports for rigid metal conduit of 32 mm in a horizontal run is

Revision exercise three

(a) 5.0 m
(b) 4.0 m
(c) 3.0 m
(d) 2.0 m

458 In the IEE Wiring Regulations, a 'double set' is equivalent to
(a) one 90° bend
(b) two 90° bends
(c) one and one-half 90° bends
(d) 3 m of straight section

459 Good earth continuity implies
(a) high resistance
(b) low resistance
(c) open circuit
(d) low conductivity

460 All cables of a.c. circuits should be contained within the same metal enclosure to avoid
(a) eddy current flow
(b) circuit overloading
(c) excessive voltage drop
(d) wiring difficulty

461 Between conductors, LOW VOLTAGE d.c. should not exceed
(a) 1500 V
(b) 1000 V
(c) 900 V
(d) 600 V

462 A PEN conductor is used throughout in a
(a) TT system
(b) TN-S system
(c) TN-C-S system
(d) TN-C system

463 While non-statutory, the 15th Edition of the Wiring Regulations complies with the fundamental requirements of the
(a) Model Construction Requirements for Petroleum Spirit Filling Stations (1984)
(b) Electricity Supply Regulations (1988)
(c) British Standards C.P. 1004 (1967) – Street Lighting
(d) British Standards C.P. 1013 (1965) – Earthing

464 The *purpose* of the IEE Wiring Regulations is to provide
(a) for every circumstance when carrying out electrical work
(b) specific details in any written electrical contract
(c) safety, especially from fire, shock, burns and injury
(d) specific instructions for untrained persons

465 The assessed current loading for six 240 V, 40 W fluorescent luminaires and their control gear is
(a) 6.0 A
(b) 5.0 A
(c) 1.8 A
(d) 1.0 A

466 The external impedance (Z_E) of a system is 0.6 Ω while maximum impedance (Z_S) for the operation of a protective device is 3.0 Ω. What is the value of R1 and R2 if the protected circuit is 24 m in length?
(a) both statements (1) and (2) are correct
(b) both statements (1) and (2) are incorrect
(c) only statement (1) is correct
(d) only statement (2) is correct

467 The British Standards code number for miniature circuit breakers is
(a) BS 1361
(b) BS 1362
(c) BS 3036
(d) BS 3871

468 The protective device of a radiant heater circuit in a bathroom must disconnect within
(a) 5.0 s
(b) 4.0 s
(c) 30 ms
(d) 400 ms

469 Assuming a domestic dwelling, the assessed current loading of a 60–A cooker circuit incorporating a 13-A socket-outlet in its control unit is
(a) 60 A
(b) 50 A
(c) 47 A
(d) 30 A

470 Which of the following does not relate to the *assessment of general characteristics* as described in the IEE Wiring Regulations?
(a) diversity
(b) earthing arrangements
(c) maximum demand
(d) testing

471 Which of the following is most likely to be a short circuit in a 3-phase, 4-wire TN-S system?
(a) blue phase to earthing conductor
(b) red phase to neutral conductor
(c) yellow phase to protective conductor
(d) neutral to earthing conductor

472 For household premises the floor area served by a single 30-A radial circuit using BS 1363 socket outlets is not to exceed
(a) 25 m^2
(b) 30 m^2
(c) 50 m^2
(d) 100 m^2

473 All the following are methods of protection against *direct contact* EXCEPT
(a) placing out of reach
(b) earthed equipotential bonding
(c) insulation of live parts
(d) provision of obstacles

474 Which of the following is not suitable for isolation purposes?
(a) circuit breaker
(b) limit switch
(c) plug and socket
(d) fuse link

475 Which of the following is not acceptable as a method of protection for a TT System?
(a) time-lag device
(b) residual current device
(c) fault-voltage operated device
(d) overcurrent protective device

476 Fixed equipment, in general, must be protected by devices which disconnect the circuit within
(a) 5.0 s
(b) 4.0 s
(c) 0.5 s
(d) 0.4 s

477 The value of 'k' for an aluminium protective conductor within a PVC/PVC cable is
(a) 125
(b) 95
(c) 81
(d) 76

478 It is recommended that drainage facilities be provided where electrical equipment contains flammable dielectric liquid in excess of
(a) 100 litres
(b) 50 litres
(c) 25 litres
(d) 5 litres

479 A *fire barrier* is a method of protection against thermal effects (Statement 1) AND may be placed between floors of a multi-storey building in a wiring system (Statement 2).
(a) both statements are correct
(b) both statements are incorrect
(c) only statement 1 is correct
(d) only statement 2 is correct

480 A guard is required on fixed equipment where the temperature is likely to exceed
(a) 100°C
(b) 80°C
(c) 60°C
(d) 30°C
Note: this requirement is relaxed if the equipment carries a BS number which admits a temperature above the answer required.

481 A bell circuit used for telecommunications is classified as a
(a) category 1 circuit
(b) category 2 circuit
(c) category 3 circuit
(d) category 4 circuit

482 A certain cable has a design current of 8 A and a volt drop of 15 mV/A/m. If it is allowed maximum volt drop (2.5%), the longest length which can be used on a 240 V supply is
(a) 100 m
(b) 75 m
(c) 50 m
(d) 25 m

483 When using the time/current characteristics in the Tables of Appendix 8 (IEE Regulations), the log term 10^0 is an expression of
(a) 100
(b) 10
(c) 1
(d) 0

484 In Table 41A2 of the IEE Regulations, a 30-A BS 1361 fuse has a maximum impedance of 2.0 Ω. For a supply of 240 V, this means that: (1) the device is limited to passing four times the circuit current and (2) the circuit current will disconnect the fuse in 0.4 s.
(a) both statements (1) and (2) are correct
(b) both statements (1) and (2) are incorrect
(c) only statement (1) is correct
(d) only statement (2) is correct

485 Which of the following is NOT recognized as suitable for acting as an effective earth electrode?
(a) non-corrosive metallic covering of cables
(b) electrodes embedded in foundations
(c) metal gas and water pipes
(d) metallic reinforcement in concrete

486 Supplementary bonding conductors having mechanical protection are subject to a minimum size of
(a) 1.5 mm^2
(b) 2.5 mm^2
(c) 4.0 mm^2
(d) 6.0 mm^2

487 In a TN-S system, a 13-A socket outlet is wired in a garage where there is an intention to supply equipment used outside the equipotential zone of the installation. The socket outlet must be protected by a residual current device designed to operate at
(a) 0.03 A
(b) 0.3 A
(c) 3 A
(d) 13 A

488 For a completed electrical installation, the minimum insulation resistance test figure is
(a) 1 MΩ
(b) 500 Ω
(c) 100 Ω
(d) 0.5 Ω

489 The assessed demand for lighting in a guest house is based on a diversity allowance of
(a) 90%
(b) 75%
(c) 66%
(d) 50%

490 The maximum horizontal distance between supports for a 20 mm rigid metal conduit is
(a) 2.25 m
(b) 2.00 m
(c) 1.75 m
(d) 1.00 m

491 The most suitable size of BS 3871 protective device for protecting a twin armoured PVC cable feeding a fixed load of 6 kW/240 V is
(a) 10 A
(b) 15 A
(c) 20 A
(d) 30 A

492 In Question 491, if the ambient temperature was 60°C, the cable selected would be
(a) 2.5 mm^2
(b) 4.0 mm^2
(c) 6.0 mm^2
(d) 10.0 mm^2

493 Referring to Questions 491–492, if the length of run was 35 m, the cable volt drop would be approximately
(a) 3.6 V
(b) 3.2 V
(c) 2.9 V
(d) 1.6 V

494 Which of the following is correct for conductor sizing?
(a) $I_B \leq I_n \leq I_Z$
(b) $I_B < I_n < I_Z$
(c) $I_B < I_n > I_Z$
(d) $I_B \geq I_n \leq I_Z$

Note: $\leq$ means equal to or less than
$\geq$ means equal or greater than
$>$ means greater than
$<$ means less than

495 If the earth fault loop impedance of a ring circuit was 0.5 Ω, and the fault current 480 A, the supply voltage would be in the range of
(a) extra-low voltage
(b) low voltage
(c) medium voltage
(d) high voltage

496 Given $S = \dfrac{\sqrt{I^2 t}}{k}$, what is the value of S when I = 400 A, t = 0.035 s, and k = 115?
(a) 1.00 mm²
(b) 0.82 mm²
(c) 0.75 mm²
(d) 0.65 mm²

497 Which of the following is not allowed in relation to reach from a person using a bath or shower?
(a) BS 3456 shower heater
(b) BS 3052 shaver unit
(c) Cord-operated switches
(d) Standard BS 1363 socket outlet

498 When testing the continuity of a protective conductor using a.c., the test voltage must not exceed
(a) 240 V
(b) 110 V
(c) 64 V
(d) 50 V

499 An impedance test of a final circuit revealed a value of 0.68 Ω. If the value of Z_E was given as 0.3 Ω, the circuit phase and protective conductor resistances would be limited to
(a) 2.04 Ω
(b) 1.66 Ω
(c) 0.98 Ω
(d) 0.38 Ω

500 The correction factor associated with a BS 3036 protective device is
(a) 1.333
(b) 0.725
(c) 0.500
(d) 0.030

Answers and comments to multiple-choice questions

Electrical installation work Part I

Students may find it useful to refer to the comments section.

1	(c)	28	(b)	55	(a)	82	(b)
2	(d)	29	(c)	56	(b)	83	(b)
3	(a)	30	(d)	57	(d)	84	(b)
4	(a)	31	(b)	58	(c)	85	(c)
5	(b)	32	(d)	59	(d)	86	(c)
6	(a)	33	(a)	60	(a)	87	(a)
7	(a)	34	(a)	61	(c)	88	(d)
8	(a)	35	(b)	62	(b)	89	(b)
9	(d)	36	(c)	63	(b)	90	(d)
10	(c)	37	(c)	64	(b)	91	(d)
11	(d)	38	(c)	65	(c)	92	(d)
12	(d)	39	(d)	66	(b)	93	(a)
13	(b)	40	(c)	67	(a)	94	(c)
14	(a)	41	(d)	68	(b)	95	(c)
15	(d)	42	(c)	69	(a)	96	(a)
16	(d)	43	(c)	70	(b)	97	(c)
17	(a)	44	(d)	71	(a)	98	(b)
18	(c)	45	(b)	72	(a)	99	(d)
19	(c)	46	(c)	73	(a)	100	(a)
20	(b)	47	(c)	74	(c)		
21	(d)	48	(b)	75	(c)		
22	(a)	49	(b)	76	(b)		
23	(d)	50	(b)	77	(d)		
24	(a)	51	(b)	78	(d)		
25	(a)	52	(d)	79	(a)		
26	(b)	53	(a)	80	(b)		
27	(b)	54	(b)	81	(a)		

Comments

1. (a) gives power, (b) gives energy and (d) gives voltage
2. (a) is the unit of p.d. or e.m.f., (b) current and (c) power
3. (b), (c) and (d) are not recognized as fuse symbols
4. (b), (c) and (d) are all low voltages
5. (a) is a two-way switch, (c) intermediate switch and (d) obsolete intermediate switch symbol
6. (b) is an ordinary tee box, (c) a branch three-way box and (d) a back-outlet through box
7. (b), (c) and (d) are made up responses
8. (b), (c) and (d) are made up responses
9. (a) is used between busbars and main switchgear, (b) between two-way switches and (c) used for supplying apparatus beyond normal wiring
10. (a), (b) and (d) are not applicable
11. (a), (b) and (c) are all possible actions but (d) is more important under the circumstances
12. (a) and (b) may be necessary as well as (c) for small burns: water will neutralize the acid
13. (a) would probably give no shock and (b), (c) and (d) different levels of shock. The key response would involve the highest voltage
14. (b) and (c) would not be used because of their conducting nature and (d) may give off toxic fumes
15. The minimum height is one metre
16. A double-insulated power tool requires no earthing
17. (a) is the correct response because of the increased use of plastic water mains and insulated pipe joints
18. Excess current implies that a circuit is being overloaded, hence (c) is the correct response
19. Response (c) is the ideal ratio
20. A first-aider should be skilled enough to check the patient's pupils and carotid pulse. If assistance was available, (d) could also be correct
21. (d) is correct. The choke also provides the initial voltage surge to start the discharge lamp but suffers from having a poor power factor
22. The potential differences (V_2 and V_3) are to be subtracted from the supply volts (V_s)
23. The voltage coil needs to be connected to the neutral of the supply
24. The answer is found by $t = \dfrac{Q}{I}$ seconds
25. The answer is found by $I = \sqrt{\dfrac{P}{R}}$ amperes
26. (a) possesses resistance, (c) possesses inductance (self) and (d) possesses capacitance
27. The answer is found by $P = \dfrac{V^2}{R}$ watts
28. (a) and (c) will cause a short-circuit when P is pressed; (d) will not stay on
29. (c) is the correct response since
   $$I = \dfrac{P}{V} \text{ amperes}$$
30. (a) is the *maximum* value, (b) the *mean* value and (c) any value or point in the a.c. wave-shape – (d) the correct response is often called the effective value
31. Switch B shorts-out the internal connections of both A and C switches, keeping the lamp alight
32. (d) the correct response. Potential difference is the only factor not shown in Figure 122
33. See IEE Regulations, Part 2: Definitions
34. $Z = \dfrac{E - V}{I} = \dfrac{240 - 235}{25} = 0.2\ \Omega$

    $I_{SC} = \dfrac{V}{Z} = \dfrac{240}{0.2}$

    $= 1200\ \text{A}$
35. Having minimal neutral current flowing implies that the phase conductors are carrying nearly equal load currents, making full use of cable size and switchgear size ratings
36. The answer is (c), found by adding up the separate R, Y and B phase loads
37. (a) and (b) are not the reason nor is (d) but the latter may in some instances be right
38. (c) is the right response with S3 and S4 giving 8 ohms, S5 and S6 giving ohms
39. (d) is the correct response and it allows the lamp to operate at a higher temperature
40. (a) is the ratio for finding efficiency

Answers and Comments to Multiple-choice Questions

41 Inductance is the chief property of a choke ballast and creates a lagging power factor. Capacitors neutralize this effect by improving the power factor
42 (c) is the correct response being expressed as $2\pi fL$ or X_L, its unit is the ohm
43 The current is found by $I = \dfrac{VA}{V}$ amperes
44 The only two likely positions are 1 and 4. If the lamp was wired in circle 1 it may jeopardize the operation of the coil if the lamp filament ruptured
45 (b) is the correct response – pushing S2 diverts the alarm signal
46 (a) and (b) are both found as electrolytes in simple primary cells and (d) in a mercury cell or alkaline cell
47 (c) is the correct response being about 1.15 to 1.2 depending on type
48 See recommended distance in the IEE Regulations, requirement 476–20
49 See the IEE Regulations, requirement 471–34
50 (b) is the correct response because this device is often found on cooking appliances
51 At 20°C (b) is the lowest at 0.0172 $\mu\Omega$m
52 (a), (b) and (c) may become affected by heat
53 PVC hardens at very low temperatures and is liable to split or crack when handled
54 (a), (c) and (d) are not related to a conductor's dimensions
55 (b), (c) and (d) are non-ferrous metals
56 (a) is the part that knocks against the bell and (c) is the name of a piece of metal which bridges the poles of a magnet to retain its magnetism (not associated with Figure 129)
57 (a) is the process of forming ions, (b) is the state of being condensed and (c) is concerned with the action of sulphur
58 (a), (b) and (c) may deteriorate with age and are substances that can burn
59 Insulation resistance is inversely proportional to length
60 (b), (c) and (d) have positive temperature coefficients of resistance
61 This is the part that keeps the die straight along the conduit
62 (b) is correct. Footprints could be used to tighten up the glands later
63 The answer is found by adding
    15 + 25 + 15 = 55 mm
64 The correct names for these products are gravity toggles and spring toggles
65 A plumb and bob line can only be used for (c)
66 A motor by itself is not a tool
67 (a) is the only tool likely to suffer
68 (b) is the only possible answer
69 See the IEE Regulations, requirement 529–5
70 Resin is a non-corrosive flux
71 See BS Codes of Practice (CP 1003: Part 1: 1964)
72 Item 1 is a distribution board, item 2 a double-wound transformer, item 4 an indicator panel and item 5 a bell
73 (a) is correct because it isolates both poles of the supply simultaneously
74 (c) is correct because it ensures that a complete order has been delivered
75 See IEE Regulations, requirement 522–6
76 (a), (c) and (d) are made up distractors
77 Bonding offers protection against indirect contact
78 (a), (b) and (c) are made up ratios sharing common factors
79 See IEE Regulations, requirement 553–15
80 See IEE Regulations, requirement 522–6
81 See the IEE Regulations, Part 2 Definitions; (a) is correct
82 See the IEE Regulations, requirement 523–6
83 Conduit sizes are determined by their outside diameter
84 See the IEE Regulations, Section 514, requirement 514–5(a)
85 Actual volt drop is the product *length, load current* and *multivolt drop per ampere metre*
86 The assessed current is found by
    10 + (0.3 × 15) + 5 = 19.5 A. See IEE Regulations, Table 4B
87 A crampet is a metal conduit fitting
88 See IEE Regulations, requirement 523–20
89 (b) is the correct response but also see the IEE Regulations, Chapter 11, Scope
90 See IEE Regulations, Part 2: Definitions and section 525
91 The ohmmeter will indicate a low resistance with the appliance switch closed
92 See the IEE Regulations, requirement 613–5

93   Impedance is the total opposition to current flow in a.c. circuits
94   See IEE Regulations, Appendix 15
95   See IEE Regulations, Table 52B
96   A *dead-short* is an extremely low ohmic reading, indicated by the tester showing zero
97   See the IEE Regulations, requirement 613–6
98   (b) is the correct response, (a), (c) and (d) are all similar in meaning
99   (a), (b) and (c) can be used to find the working resistance of a circuit or apparatus but two would need connection in circuit
100 See the IEE Regulations, Appendix 16

Note: Reference should be made to the IEE Regulations for Electrical Installations

Electrical installation work Part II

Students may find it useful to refer to the comments section.

101	(d)	121	(c)	141	(c)	161	(c)	181	(d)
102	(a)	122	(b)	142	(a)	162	(d)	182	(d)
103	(b)	123	(c)	143	(b)	163	(a)	183	(c)
104	(a)	124	(a)	144	(a)	164	(b)	184	(b)
105	(a)	125	(d)	145	(a)	165	(a)	185	(c)
106	(a)	126	(a)	146	(b)	166	(c)	186	(b)
107	(b)	127	(c)	147	(d)	167	(a)	187	(a)
108	(c)	128	(d)	148	(c)	168	(b)	188	(b)
109	(d)	129	(a)	149	(a)	169	(c)	189	(a)
110	(c)	130	(a)	150	(c)	170	(d)	190	(a)
111	(c)	131	(d)	151	(d)	171	(d)	191	(c)
112	(d)	132	(a)	152	(d)	172	(a)	192	(a)
113	(c)	133	(c)	153	(b)	173	(c)	193	(c)
114	(a)	134	(a)	154	(b)	174	(c)	194	(a)
115	(a)	135	(b)	155	(a)	175	(c)	195	(c)
116	(c)	136	(c)	156	(b)	176	(a)	196	(d)
117	(a)	137	(b)	157	(b)	177	(c)	197	(b)
118	(b)	138	(d)	158	(b)	178	(c)	198	(c)
119	(a)	139	(c)	159	(d)	179	(b)	199	(d)
120	(a)	140	(c)	160	(c)	180	(b)	200	(d)

Comments

101 This is Regulation 1 of the Electricity (Factories Act) Special Regulations
102 (a) is the correct response and should show the most economical method of using labour
103 (a), (c) and (d) are concerned with site organization
104 There are two bodies that look after the interests of craftsmen and apprentices in the electrical contracting industry: the JIB and the EETPU
105 This is a method of tendering and involves all the materials and labour required for a contract

Answers and Comments to Multiple-choice Questions

106 (b) is responsible for running a whole contract and (c) and (d) are both agents of the architect offering a specialist service
107 (b) is the correct response being marked on site as the installation progresses
108 (c) is the correct response which requires specific requirements for wiring an installation as well as incorporating standards of workmanship in accordance with the IEE Wiring Regulations
109 See BS 3939 graphical symbols
110 See *Electrical Installation Technology 1*, page 98
111 (a) and (b) refer to testing and inspection of an installation and (d) is an architect's instruction
112 (d) is the correct response. See IEE Regulations, requirement 537–13
113 (a), (b) and (d) are too general. See Regulation 12 of the Electricity (Factories Act) Special Regulations 1908 and 1944
114 See BS 5345. There is no zone 3
115 (a) is the correct response. See IEE Regulations, requirement 547–2
116 The answer is (c). This is to avoid all possible danger
117 See the IEE Regulations, Table 11B
118 (b) is the correct response but also see the IEE Regulations, requirement 461–4
119 See the IEE Regulations, requirement 523–23
120 For maximum earth-loop impedances see the IEE Regulations, requirement 413–5
121 The answer is found by

$$V_P = \frac{V_L}{\sqrt{3}} = \frac{110}{1.732} = 64 \text{ V}$$

122 (b) is the non-standard value
123 The *phase* current is the *line* current
124 The answer is found by

$$V_L = \sqrt{3} \times 240 = 415 \text{ V}$$

125 (d) is the correct response since the system is balanced
126 The system would become unbalanced causing excess current to flow in two phases
127 (c) is correct, (d) is for a star connected system
128 (d) is correct showing link between neutral and earth
129 The total power is the addition of the two wattmeter readings
130 $\cos \phi = \dfrac{\text{adjacent}}{\text{hypotenuse}} = \dfrac{\text{kW}}{\text{kVA}}$

but $\text{kW} = \sqrt{\text{kVA}^2 - \text{kVA}_r^2}$

$= \sqrt{100 - 35} = 8 \text{ kW}$

thus $\cos \phi = \dfrac{8}{10} = 0.8$

131 (a) has no fuses, neither has (b) and in (c) the fuses are part of the moving assembly of the switch
132 (a) is the correct response. See manufacturer's catalogues or Appendix 8 Regulations
133 (c) is correct since other fuses in the system are required to remain healthy
134 See IEE Regulations, requirement 523–5
135 The resistance is found by $R = \sqrt{Z^2 - X^2}$

and power factor by $\dfrac{R}{Z}$

136 A residual current device will not operate unless the fault detector coil senses an out of balance flux in the ring. This will not happen in (a), (b) or (d)
137 See *equipotential bonding* in the IEE Regulations, Part 2
138 The answer is (d), given in the IEE Regulations
139 (c) is correct (see the IEE Regulations). (a) is the ratio of average load and maximum load, (b) allows for future loading and (d) is used in cable selection procedures as a correction factor
140 While a semiconductor thyristor is a switching device, it cannot be used for circuit isolation. See the IEE Regulations, requirement 537–4
141 The answer is (c) mineral-insulated metal-sheath cable
142 When the compression ring is squeezed inside the gland by the back nut, it provides electrical continuity against the cable's armouring, it does not provide mechanical protection
143 MIMS is ideal for installations with above normal temperatures
144 Links are required where lengths and fittings are joined together to provide electrical earth continuity

145 Eddy currents are created by a single a.c. conductor's alternating magnetic field cutting through the metal conduit
146 (b) is correct. The system is restricted in use (see IEE Regulations)
147 The moisture must be dried out. If the end was re-made the problem may still exist
148 See British Standard CP 1003
149 This is a method of circulating current protection used for portable tools and transportable equipment. See BS Codes of Practice (CP 1013: 1965)
150 Statement 2 is incorrect – permissible volt drop must not exceed 2.5% of the declared voltage. See IEE Regulations, requirement 522–8
151 (a), (b) and (c) are all d.c. motors, (b) can be operated on a.c.
152 (d) consists of copper or aluminium bars shorted out at each end by metal end rings. (b) has a commutator and (a) and (c) are the same
153 Changing two supply lines changes the sequence of the three-phase rotating magnetic field produced by the stator windings
154 (b) is correct
155 (a) is correct
156 (b) is correct. Speed being inversely proportional to flux means that weakening the field system (by increasing field resistance) the motor will run faster
157 Motors identified with this coding are classified as non-sparking and are used in hazardous locations
158 (b) is correct since induction motors have poor power factors
159 (a), (b) and (c) are the important factors
160 (c) is correct because changing the supply leads over changes both the polarity of the main field and the armature field. Statement 2 does not apply
161 The rotor speed is found by $n_r = n_s (1 - s)$
162 The calculation of the hoist output is

$$P = \frac{\text{energy (N-m)}}{\text{time (s)}} \text{ watts}$$

The motor's output (hoist's input)

$$= \frac{\text{hoist's output}}{\text{efficiency}} \text{ watts}$$

163 Percentage efficiency $= \dfrac{\text{output}}{\text{input}} \times 100$

$$= \frac{373\,000}{4200}$$

$$= 88.8\%$$

164 (b) is correct. Use should be made of a straight edge parallel to the shafts
165 Total power is $3V_p I_p \cos \phi$
166 The answer is found by $\dfrac{\text{phase voltage}}{\text{line voltage}} \times 100$
167 See *Electrical Installation Technology 2*, Fig. 4.6
168 See *Electrical Installation Technology 2*, Fig. 4.10
169 There is no electrical connection between stator and rotor
170 An open circuit field winding would not cause blackening on a commutator
171 (d) is correct, its unit being the lux
172 (a) is correct, it allows for losses as a result of absorption of light by walls and floor as well as ceiling and light fittings, it is always less than unity
173 (a) is illuminance, (b) is luminous flux and (d) luminance
174 (c) is the answer, based on the inverse square law

$$(E = \frac{I}{d^2})$$

175 (c) is the answer, based on the cosine law

$$(E = \frac{I}{d^2} \cos \phi)$$

Note: The distance in question 174 is different from the distance in question 175

176 (a) is the correct response, the distractors are made up
177 Stroboscopic effect is peculiar to discharge lighting when connected to the a.c. supply. As the supply passes through zero twice every cycle, the light produced tends to flicker at twice mains frequency
178 (c) is correct, being a factor which considers a circuit power factor of not less than 0.85 as well as control gear losses and harmonics

Answers and Comments to Multiple-choice Questions 189

179 Figure 53 is used to start a fluorescent tube
180 See the IEE Regulations, requirement 554–3
181 (d) is correct. See BS 3939 graphical symbols
182 (a) and (b) allows one lamp to operate and (c) will not allow any lamp to operate
183 (c) is the correct response, (a) is forward biased
184 (a) is half-wave rectification, (c) is sinusoidal and (d) is a square wave
185 The gate needs to be triggered
186 (a) is the clearance between stator and rotor in a motor, (c) is formed by the interchange of valence electrons between atoms and (d) is an atom which has either gained or lost an electron
187 (a) is correct, (b) is a positive terminal, (c) is a negative terminal and (d) is the connection region between emitter and collector in a transitive device
188 (a) would be correct it it was non-linear, (c) is used to carry a lightning discharge down to earth, and (d) senses earth-leakage currents
189 The diode needs to be by-passed and the gate of the thyristor pulsed
190 (a) is correct, being otherwise called a smoothing circuit
191 (c) is correct which means that high ohmic values are sought
192 See the IEE Regulations, requirement 614
193 See the IEE Regulations, Table 52B
194 The shorter the length of cable the better its insulation resistance will be
195 (c) is correct, otherwise the second auxiliary electrode will not give a constant reading
196 See the IEE Regulations, Appendix 16
197 (b) is correct since only a small current is required to produce full scale deflection on the instrument
198 See *Electrical Installation Technology 2*, Fig. 5.7, page 9
199 The ammeter will read 2-A
200 Insulation resistance is the same as connecting resistors in parallel, the answer will be less than the smallest value recorded, i.e. response (d)

Note: Reference should be made to the IEE Regulations for Electrical Installations

Revision exercise one

201	(c)	221	(d)	241	(a)	261	(c)	281	(c)
202	(b)	222	(d)	242	(c)	262	(c)	282	(b)
203	(d)	223	(d)	243	(d)	263	(c)	283	(d)
204	(a)	224	(b)	244	(b)	264	(b)	284	(a)
205	(d)	225	(d)	245	(c)	265	(a)	285	(a)
206	(d)	226	(c)	246	(c)	266	(c)	286	(c)
207	(c)	227	(c)	247	(d)	267	(c)	287	(b)
208	(d)	228	(c)	248	(a)	268	(a)	288	(a)
209	(b)	229	(c)	249	(c)	269	(a)	289	(a)
210	(b)	230	(d)	250	(a)	270	(d)	290	(d)
211	(a)	231	(c)	251	(c)	271	(a)	291	(b)
212	(c)	232	(b)	252	(a)	272	(a)	292	(d)
213	(a)	233	(d)	253	(b)	273	(a)	293	(b)
214	(b)	234	(c)	254	(c)	274	(d)	294	(c)
215	(c)	235	(a)	255	(d)	275	(b)	295	(b)
216	(d)	236	(b)	256	(c)	276	(b)	296	(c)
217	(a)	237	(a)	257	(d)	277	(b)	297	(d)
218	(c)	238	(c)	258	(c)	278	(a)	298	(c)
219	(b)	239	(d)	259	(a)	279	(a)	299	(b)
220	(b)	240	(a)	260	(a)	280	(b)	300	(c)

Comments

201 See Table 41A2 (b) IEE Wiring Regulations
202 See BS 3939 Symbols in EIT V1*
203 (d) is correct, it is an integrating meter
204 (a) is correct, see EIT V2 for names of units
205 (d) is correct, see EIT V1, Figure 83, page 94
206 (d) is correct since it is used to stop cable insulation becoming damaged when drawn into conduit
207 See page 11 EIT V2
208 (d) is correct – count nearest digit behind dial arrow noting direction of rotation
209 See page 91 EIT V1
210 See page 10 EIT V2
211 See part 2 IEE Wiring Regulations
212 (a), (b) and (d) are methods of finding the value of an unknown resistor
213 The resistance path through the circuit is very low when all switches are closed; most current passing through S1, S2 and S4 – L1 and L2 will not appear to light
214 (b) is correct – re-draw circuit without switches
215 See page 10 EIT V2
216 See page 10 EIT V2
217 (a) is the only tool listed designed for the purpose
218 (c) is correct
219 (a) is used in conjunction with a d.c. voltmeter, (c) is a variable resistor not used for the purpose required and (d) is a variable transformer
220 Whilst (a) and (c) oppose current flow, in an a.c. circuit impedance is the overall opposition. (d) is the specific resistance of a unit cube of material
221 (d) is correct, see page 80 EIT V1
222 (d) is correct, S2 and S3 are intermediate switches
223 (d) is correct. S1 shorts out the 48 Ω resistor, S2 allows 23 Ω to be the parallel resistance value and S3 closed gives 23 + 52 = 75 Ω

224 (b) is correct – during discharge lead peroxide and porous lead are slowly changed into lead sulphate
225 (d) is correct when all cells are connected in series
226 The answer is (c), found in the IEE Wiring Regulations, Appendix 3, Figure 3
227 Add both resistance values and divide into the cell voltage
228 (a) is the product of power and time, (b) is heat energy and (d) is the energy of a body because of its position
229 The question is aimed at finding power factor (c) using all the instruments, i.e.

$$\text{P.F.} = \frac{\text{Power (wattmeter)}}{\text{Voltamperes (voltmeter and ammeter)}}$$

230 (d) is the answer since the other values given are too high
231 The answer (c) is found from the expression $\frac{2.75}{110} \times 100$
232 Fusing factor is mentioned on page 57 of EIT V1. The answer is (b) being the product of 1.2 and 5
233 The transformer has a step-down ratio and therefore (d) is the correct response – see pages 90–93 of EIT V2
234 Power (P) is proportional to voltage $(V)^2$. By reducing the voltage to one-half its original value the power consumed is reduced to one-quarter, i.e.

$$P \propto \left(\frac{1}{2}\right)^2 = \frac{1}{4}$$

235 The plus is the upper limit and (a) is correct, i.e.

$$\frac{5}{100} \times 4000 = 200,$$

therefore 200 + 4000 = 4200 Ω

236 See page 79 EIT V1
237 See page 63 EIT V2
238 (c) is the answer since work done is force times distance

---

*EIT V1, V2 and V3 are abbreviations for Electrical Installation Technology volumes 1, 2 and 3 written by the same author.

Answers and Comments to Multiple-choice Questions

239 Reference should be made to Table 5A, Appendix 5, IEE Regulations
240 See Table 52B, IEE Regulations
241 See page 57 EIT V1
242 (c) is the correct response
243 See Regulation 553–19 IEE Regulations
244 The answer is found from

$$\frac{2.5}{100} \times 415 = 10.375 \text{ V}$$

245 See comments given in Question 53
246 (c) is the answer, brass has a positive temperature coefficient of resistance
247 In Figure 75, if the voltmeter were taken out of the circuit and replaced by an ammeter, no current would flow. This is because both batteries are connected to produce equal and opposite potentials to each other, thus (d) is correct
248 (a) is the correct response. The concept is to invent rules to enable the principles to be understood. Current flowing inwards (noted by a cross) creates a clockwise magnetic field whilst current flowing outwards (noted by a dot) creates an anticlockwise magnetic field. Equal and opposite fields will tend to repel each other
249 See page 15 EIT V1
250 Volume has three dimensions. Multiply the area $\frac{\pi d^2}{4}$ by the length (l).

Remember that the diameter is in millimetres and 1 m = 1000 mm
251 The abbreviation r.m.s means root mean square and is the *effective value* of an a.c. current or voltage equivalent to the same value of d.c. current or voltage
252 (a) is correct. (b) exerts a downwards force of 9.81 N on a mass of 1 kg, (c) is the force which holds things together and (d) is the force created by a magnetic field, e.g. electromagnet
253 (a) is the capacity for doing work, (b) is correct, (c) is the cause of motion and (d) is associated with the rate at which velocity changes with respect to time
254 Normal lighting circuits are wired in parallel (c)
255 The lamp shown is a low-pressure sodium vapour discharge lamp
256 (c) is correct
257 A four-wire system will have a neutral to carry any out of balance phase currents
258 The formula for c.s.a is given in Question 250 above
259 See IEE Regulations, requirement 529–5
260 (a) is the answer since the system becomes unbalanced
261 See IEE Regulations, requirement 522–8
262 Not only is a 'live' feed required but also a 'live' neutral too
263 See page 21 EIT V1
264 The p.d. across the resistors is 84 V, therefore since $P = \frac{V^2}{R}$ power consumed is 588 W
265 (a) is the answer but also see IEE Regulations, Table 52A
266 The parallel group become 12 μF and this in series with the other 12 μF capacitor becomes 6 μF
267 Q = CV which is $6 \times 10^{-6} \times 500 = 3$ mC
268 (a) is correct
269 (a) is correct, but see also page 3, EIT V1
270 $Z = \frac{V}{I} = \frac{275\,000}{500} = 550 \, \Omega$
271 (a) is correct
272 (a) is correct, i.e. $\frac{P}{VI} = 1$
273 (a) is correct, i.e.

$$r = \frac{E - V}{I} = \frac{10 - 9}{8} 125 \text{ m}\Omega$$

274 (d) is correct (0.2 kV divided by 16)
275 The fuse will rupture
276 An alternating supply current extinguishes itself twice every cycle, i.e. 100 times on a 50 Hz supply
277 (b) is correct, one of the switch wires should be connected to terminal 2
278 (a) will corrode. Galvanized conduit would be suitable

279 1 joule = 1 wattsecond; 1 hour = 3600 seconds. Thus,
1 kWh = 1000 × 3600 = 3 600 000 = 3.6 MJ
280 See IEE Regulations, Table 5A
281 (c) is correct being low voltage
282 Time = $\dfrac{\text{Energy}}{\text{Power}} = \dfrac{5 \times 10^6}{2.5 \times 10^3} = 2$ ks
283 The resistors need to be connected in parallel
284 (a) is correct since a.c. is initially created
285 (a) is correct since fluxes allow the metals to adhere to each other
286 (c) is correct. See BS 3939 graphical symbols
287 Phase voltage = $\dfrac{\text{Line voltage}}{1.732}$
288 (a) is the answer since it is a visual inspection that is being made
289 See IEE Regulations, requirement 413–6
290 (d) is correct (see atomic weight table for elements)
291 (b) is correct (4 × 2.5 mm² circuit conductors and two circuit protective conductors)
292 (d) is correct (see 'Spurs' page 118, IEE Regulations)
293 See IEE Regulations, requirement 547–6
294 (c) is correct being 40% tin and 60% lead
295 (b) is correct being the standard practice using Tables in Appendix 9, IEE Regulations relating to current carrying capacities and voltage drop
296 PVC covering on MIMS cable provides sufficient protection against corrosive environments and dampness
297 (d) is correct, shorting out the poles with a 'keeper' is the normal procedure allowing the magnet to retain its magnetism
298 (c) would be the correct response. Phasor diagrams are rotating a.c. quantities and in Figure 197 the current lags the voltage by some phase angle (see page 57, EIT V2)
299 See *exclusions from scope*, IEE Regulations, page 2
300 See IEE Regulations, requirement 471–34

Revision exercise two

301	(d)	321	(b)	341	(d)	361	(b)	381	(a)
302	(b)	322	(d)	342	(c)	362	(d)	382	(a)
303	(a)	323	(c)	343	(d)	363	(d)	383	(a)
304	(d)	324	(d)	344	(d)	364	(b)	384	(b)
305	(a)	325	(a)	345	(a)	365	(b)	385	(a)
306	(d)	326	(d)	346	(c)	366	(d)	386	(d)
307	(c)	327	(d)	347	(b)	367	(a)	387	(a)
308	(b)	328	(b)	348	(c)	368	(a)	388	(a)
309	(a)	329	(c)	349	(b)	369	(c)	389	(c)
310	(b)	330	(d)	350	(a)	370	(d)	390	(d)
311	(b)	331	(a)	351	(d)	371	(b)	391	(d)
312	(c)	332	(c)	352	(b)	372	(b)	392	(a)
313	(a)	333	(b)	353	(a)	373	(c)	393	(c)
314	(d)	334	(c)	354	(c)	374	(c)	394	(b)
315	(a)	335	(a)	355	(d)	375	(a)	395	(b)
316	(a)	336	(d)	356	(b)	376	(d)	396	(d)
317	(a)	337	(b)	357	(c)	377	(a)	397	(b)
318	(b)	338	(a)	358	(a)	378	(d)	398	(c)
319	(b)	339	(a)	359	(a)	379	(b)	399	(b)
320	(c)	340	(c)	360	(c)	380	(b)	400	(d)

Answers and Comments to Multiple-choice Questions

Comments

301 Take leading pulse gate line (2) as start. At (4) the wave becomes extinguished and starts again at (6) being further extinguished at (8)

302 (a) is the main supply terminals, (c) is made up and (d) is made up

303 (a) is correct but see page 95, EIT V2

304 See page 92, EIT V1

305 See page 65, EIT V1

306 It must be winding connections that have to be changed

307 Consider a phase winding to have an impedance of 10 Ω, then in *star* connection,
$I_L = I_p = V_p/Z = 240/10 = 24$ A
In *delta*, $I_p = V_l/Z = 415/10 = 41.5$ A, but
$I_L = \sqrt{3} \times I_p = \sqrt{3} \times 41.5 = 71.8$ A
Thus it will be seen that in star I_L is
reduced by $\frac{24}{71.5} = \frac{1}{3}$

308 (b) is correct, see also page 165, EIT V1

309 The critical path is the longest line through the project, i.e. activities 0–1, 1–2, 2–4, 4–8, 8–12, 12–15 and 15–16. See page 000 of this book

310 Magnetic flux will enter the salient pole from the yoke making response (b) correct. See Figure 4.19, page 82 of EIT V2

311 Subtract readings and multiply by unit cost, then add fixed charge to give answer (b)

312 (c) is correct, being the ratio frequency/pole pairs – this field cuts the rotor as it travels

313 A lightly loaded induction motor has a poor power factor

314 50 Hz is the occurrence of 50 cycles in one second, thus one cycle is $1/50 = 0.02$ s

315 (a) is correct but see Figure 89, page 99, EIT V1

316 Both windings are completely separate from each other and earth

317 Growth factor is an allowance for future loading

318 (b) is correct, being a mass of metal for dissipating heat away from the semiconductor device (see Figure 100, in this book)

319 (b) is correct – it takes 4200 J to heat 1 kg and raise its temperature by 1°C

320 Efficiency = $\frac{\text{Output (Ah discharge)}}{\text{Input (Ah charge)}}$

321 (b) is correct (see page 112 of this book)

322 The supply is in r.m.s. form for
a.c., thus $V_{max} = \frac{415}{0.707} = 587$ V

323 Force (F) = BIL Newtons
$= 15 \times 0.5 \times 0.015 = 112.5$ mN

324 $V = \sqrt{3}IR = 1.732 \times 0.011 \times 500 = 9.5$ V (approximately)

325 (a) is correct (see *Model construction requirements for petroleum spirit filling stations and similar private refuelling installations* (1984))

326 See IEE Regulations, Table 11C

327 See IEE Regulations, Section 514

328 (b) is correct (see Figure 5.18, page 99, EIT V2)

329 (c) is correct (see previous comments)

330 (d) is correct

331 The insulation resistance will be ten times better than before because it is inversely proportional to length

332 This is a requirement in the IEE Regulations, Regulation 422–5

333 (b) is correct (noise level will be more noticeable)

334 (c) is correct, (a) may cause sparking, (b) not a fault term and (d) an out-of-step occurrence in synchronous motors

335 No secondary magnetic field could be set up for the motor to start

336 Testing a live conductor to earth is to determine its insulation resistance (d)

337 (b) is correct (see IEE Regulations, Tables 12C and 12D (358))

338 See page 11, EIT V1

339 The fuses in the test prods safeguard against internal faults – (a) is correct

340 See IEE Regulations, Regulation 537–19

341 An induction motor has a lagging power factor

342 25 A $+ 16$ A $+ 21$ A $= 62$ A

343 See page 81, EIT V1

344 (d) is correct (see page 95, EIT V1)

345 (a) is correct, apply 100 W/lamp (Table 4A, IEE Regulations)

346 See Section 614, IEE Regulations

347 (c) is correct (see Table 11B, IEE Regulations)

348 (c) is correct (see Definitions, IEE Regulations)

349 (b) is correct; (a) is an energy regulator, (b) is a room thermostat and (d) is a frost-stat
350 (a) is correct (see page 76, EIT V2)
351 (d) is the possible cause
352 (b) is correct
353 (a) is correct (approximately 15 lm/W)
354 Determine distance first then find illuminance
355 (a), (b) and (c) are all likely causes
356 The question is referring to visual inspection: the others are tests
357 The ammeter is wired in series with the c.t. and the voltmeter wired across the secondary winding of the v.t. – a v.t. is also called a potential transformer
358 All phase to earth faults are phase to neutral faults
359 An autotransformer has only one phase winding, i.e. if it is for single phase use
360 $E = V - I_a R_a = 220 - (20 \times 0.15) = 217$ V
361 If no resistor was connected, (a) would be correct; (c) is for unity power factor conditions, that would be if the capacitor was not connected; and (d) is typical for an inductive circuit such as a coil
362 $\text{Input} = \dfrac{\text{Output}}{\text{Efficiency}} = \dfrac{20 \times 5 \times 9.81}{12.26 \times 0.8}$

= 100W (approximately)
363 $E = \dfrac{I}{d^2} \times \cos\theta = \dfrac{160}{25} = \dfrac{4}{5} = 5.12$ lx
364 The ratio is 5:1, therefore $5 \times 20 = 100$ A in secondary winding
365 $I = \sqrt{6^2 + 8^2} = 10$ A
366 In practice a coil will always have resistance but under the conditions described both components are in anti-phase with each other and thus the current is zero – assuming capacitor still took 8 A
367 $X_c = \dfrac{1}{2\pi fC}$, thus (a) is the correct answer
368 (b) is correct since full load $kVA = \dfrac{P}{\text{p.f.}}$
369 The low value shunt resistor has to take 4.985 A and since the p.d. across it is 75 mV (15 mA × 5 Ω),

$R_{shunt} = \dfrac{0.075}{4.985} = 15$ mΩ

370 (d) is the correct answer (see lighting manufacturers' catalogues for polar curves)
371 (b) is correct (see BS 3939 graphical)
372 See IEE Regulations, Regulation 12–6
373 See Table 55A, IEE Regulations
374 See Figure 4.17, page 80, EIT V2 for method of connection.
375 See IEE Regulations, Regulations 525–3 to 525–9
376 (d) is correct
378 (a) is correct (see page 82, EIT V2)
379 30 m is the maximum length (see column 2, Table 11B, IEE Regulations)
380 See Figure 5, Notes, Appendix 3, IEE Regulations
381 (a) is correct
382 (a) is correct (see page 44, EIT V1)
383 (a) is correct (see page 80, EIT V2 ($T \propto I^2$))
384 The motor is described on page 78, EIT V2
385 (a) is correct, i.e. $X_L = X_C$
386 Output power (P_O) = $2\pi nT$ watts – no speed, no developing power
387 $I = \dfrac{V}{R} = \dfrac{250}{5} = 50$ A
388 See page 73, EIT V2 –

$s = \dfrac{25 - 23.75}{25} = 0.05$ p.u.

389 (c) is correct
389 $R = \dfrac{p \times L}{A} = \dfrac{17.8}{10^6} \times \dfrac{10^5}{2.5} = 712$ mΩ
391 $\dfrac{1}{R} = 0.025 + 0.033 + 0.5 + 1 = 1.558$

$\dfrac{R}{1} = \dfrac{1}{1.558} = 0.64$ Ω

392 (a) is the appropriate response
393 (c) is correct – specific gravity of water is 1, for a fully charged cell (lead–acid cell), the electrolyte is about 1.28
394 (b) is correct
395 (b) is correct (see page 83, EIT V1)
396 (d) is correct (see page 93, EIT V1)
397 Poor discrimination, both devices will operate together
398 (c) is correct
399 (b) is correct – a shunt motor
400 See BS CP 1013, page 127

Answers and Comments to Multiple-choice Questions

Revision exercise three

401	(d)	421	(b)	441	(b)	461	(a)	481	(b)
402	(b)	422	(d)	442	(c)	462	(d)	482	(c)
403	(a)	423	(d)	443	(b)	463	(b)	483	(c)
404	(a)	424	(b)	444	(a)	464	(c)	484	(c)
405	(d)	425	(a)	445	(b)	465	(c)	485	(c)
406	(b)	426	(b)	446	(c)	466	(d)	486	(b)
407	(b)	427	(a)	447	(b)	467	(d)	487	(a)
408	(a)	428	(b)	448	(a)	468	(d)	488	(a)
409	(b)	429	(b)	449	(d)	469	(d)	489	(b)
410	(c)	430	(a)	450	(d)	470	(d)	490	(c)
411	(b)	431	(d)	451	(a)	471	(b)	491	(d)
412	(d)	432	(a)	452	(d)	472	(c)	492	(d)
413	(c)	433	(c)	453	(b)	473	(b)	493	(a)
414	(d)	434	(c)	454	(a)	474	(b)	494	(a)
415	(c)	435	(d)	455	(c)	475	(a)	495	(b)
416	(c)	436	(d)	456	(d)	476	(a)	496	(d)
417	(b)	437	(a)	457	(d)	477	(d)	497	(d)
418	(b)	438	(d)	458	(a)	478	(c)	498	(d)
419	(a)	439	(c)	459	(b)	479	(a)	499	(d)
420	(b)	440	(a)	460	(a)	480	(b)	500	(b)

*Comments**

401 See Part 1, Chapter 13
402 See Figure 6, Appendix 3
403 Regulation 525–3
404 See Section 613
405 Regulation 522–8
406 Regulation 413–18
407 Regulation 433–2 and Section 533
408 Regulation 413–6
409 $6.25/0.025 = 250$ V
410 TNC and TN-C-S (0.35 Ω), TNS (0.8 Ω) and TT (21 Ω)
411 Regulation 434–6
412 Regulation 422–2
413 Regulation 471–34
414 See *Exclusions from scope*, Chapter 11
415 See Table 4A, Appendix 4
416 $I = \dfrac{80 \times 6}{240 \times 0.5} = 4$ A

*Reference should be made to the IEE Wiring Regulations unless otherwise indicated.

417 See BS codes of practice 5839
418 The insulation resistance will be four times greater
419 Regulation 613–5
420 Fusing factor is the ratio of minimum fusing current and current rating
421 See Section 413
422 See Definitions, Part 2
423 See Figure 5, Appendix 3
424 See Figure 17, Appendix 15
425 See Figure 6, Appendix 3
426 Table 41A2
427 $I_z = \dfrac{45}{0.725 \times 0.94} = 66$ A
428 Table 41A2
429 Table 41A1
430 Regulation 413–4 (0.4 s)
431 See Appendix 4
432 See Appendix 9
433 See Regulation 28, page 157, EIT V1
434 See Definitions, Part 2
435 See Definitions, Part 2

436 See Appendix 16
437 $V = 40 \times 30 \times 7.4 \times 0.001 = 8.88$ V
438 Table 4A, Appendix 4
439 See Table 54F
440 Regulation 547–4
441 Regulation 476–13
442 Regulation 613–8
443 Regulation 537–4
444 Regulation 412–6
445 Table 5A, Appendix 5
446 $L = \dfrac{V_{max} \times 1000}{I_B \times mV}$
447 Table 12C, Appendix 12
448 (b) is a technical agreement, (c) and (d) are codes of practice
449 Table 5A, Appendix 5
450 Regulation 413–1 – indirect contact
451 Regulation 547–3
452 Table 54F, Appendix 5
453 Regulation 533–6
454 See Definitions, Part 2
455 The most sensible thing to do is (c)
456 Transpose for I in $P = I^2R$ watts
457 Table 11C, Appendix 11
458 See Appendix 12
459 (b) is correct, approaching zero measurement
460 Regulation 521–8
461 See Definitions, Part 2
462 See Figure 3, Appendix 3
463 Regulation 12–3
464 Regulation 21–1
465 See Note at foot of Table 4A, Appendix 4
466 See Appendix 8
467 See Tables 41A1 and 41A2
468 Regulation 471–36
469 See Table 4A, Appendix 4
470 See Part 3, Chapter 31
471 See definition of short circuit current
472 See Table 5A, Appendix 5
473 Regulation 412–1

474 Regulation 537–5
475 Regulation 413–2
476 Regulation 413–4 (ii)
477 Table 54C, Chapter 54
478 Regulation 422–5
479 Regulations 523–6 and 528–1
480 Regulation 423–1
481 Regulations 525–3 and 525–5
482 See comments to Question 446
483 Each line on log–log graph paper represents a value, 10^0 (i.e. ten to the power of zero) = 1, $10^1 = 10$, and $10^2 = 100$, etc.
484 $I = \dfrac{240}{2} = 120$ A and $\dfrac{120}{30} = 4$

Statement (1) is correct. Table 41A2 is for fixed equipment to disconnect within 5 s. See Figure 10, Appendix 8

485 Regulation 542–10
486 Regulation 547–5
487 Regulation 413–9
488 Regulations 613–6 and 613–7
489 Table 4B, Appendix 4
490 Table 11C, Appendix 11
491 $I_B = \dfrac{P}{V} = \dfrac{6000}{240} = 25$ A, therefore $I_n = 30$ A

(see Table 41A2)
492 See Table 9D2, Appendix 9

$(I = \dfrac{30}{0.5} = 60$ A$)$

493 $V = L \times I_B \times mV = 35 \times 25 \times 0.0043$
     $= 3.76$ V

494 Regulation 433–2
495 $V = I \times R = 480 \times 0.5 = 240$ V, which is (b)
496 Use formula in Regulation 543–2
497 Regulation 471–34
498 See Appendix 15
499 See Appendix 8
500 Regulation 433–2

Appendices

Appendix 1

Examination Hints

1. Long before the examination date, students should be in possession of all the necessary drawing aids for use in the exams. These should include: protractor, pen and pencils, metric ruler, radius aid or set of compasses, eraser and pencil sharpener. A scientific calculator is quite acceptable in both Part I and Part II examinations. It is important to know how to use the calculator as well as any new drawing aids recently purchased before the exam.

2. Arrive at the examination centre at least 15 minutes before the start of the exam in order to find out where it is held and also to settle in.

3. On receiving the exam question book, read the instructions very carefully, finding out what is to be expected, how many questions are to be answered and what exam material/aids can be used. Pay particular attention to the time allowed to complete the exam. First written, multiple-choice question papers require 70 items to be completed in 2 hours while second written papers require 6 questions to be answered in 3 hours. (This may soon change.)

4. In multiple-choice exams, candidates are expected to have: (i) their individual timetable; (ii) question book; (iii) answer sheet; and (iv) HB pencil. The present instructions on the question book ask for candidate's name, number and signature including the Centre number. The answer sheet requires this and other information, such as date and paper number. For each question, a candidate has to fill in one of four boxes lettered 'a', 'b', 'c' and 'd' and if an error is made then the lower half of the box is completed before making another choice.

Questions should be answered by working down each column and going back to those items which were more difficult to answer. It is important to check that the answer sheet number corresponds with question paper number. It is unfortunate that many candidates walk out of the exam room before the exam ends. This is not good practice because the student has spent a whole year studying for this special day, so why give up at this stage when there is time available to thoroughly check the script! Students must remember that multiple-choice questions test knowledge over the whole range of the electrical syllabus. In Course 236 Part I Certificate, the topic areas include:

Health and Safety
Installation Circuits
Materials Commonly Used in Electrical Installation Work
Tools, Equipment and Working Processes
Measurement and Setting Out
Installation Procedures and Techniques
Inspection and Testing of Installations

In Course 236 Part II Certificate, the topic areas include:

Regulations
Health and Safety
Distribution
Consumer's Switchgear and Earthing
Wiring Systems
Installation of Machines
Rectification
Installation of Lighting
Inspection, Testing and Measurement

Note: One section of this book contains multiple-choice questions, and comments about these can be found in the answer section.

5. In second written papers (Part II only), the present instructions allow students to take their own copy of the IEE Regulations into the examination. In a recent City and Guilds report on examinations at this level, the general comments reflected that *a number of candidates either had little knowledge of the IEE Regulations or did not possess a copy of them*. This can only be seen as poor exam preparation. The IEE Regulations is an essential document on the course and whilst it is acknowledged that many students do not fully understand it, they must at least have been acquainted with its various parts and appendices in their studies.

Examiners are continually confounded by students' lack of basic electrical principles, shallow treatment of descriptive questions and poorly drawn diagrams. The instructions are quite explicit that all questions carry equal marks, so it is important that students should thoroughly read all nine questions and select the best six. A further exam instruction states that it is advisable to show all stages in calculations but far too many candidates fail miserably in this area. As a guide, examination questions can be categorized under three main headings, namely: **(i) descriptive type; (ii) calculation type;** and **(iii) drawing type**. These ways of obtaining information can often be found in one particular question so it is very important to know what examiners are looking for in each area. These are briefly summarized below.

Descriptive questions need students to express their full understanding of the subject. Answers should be thought out first before writing anything down, making reference to the practical and/or theoretical aspects of the question. Some degree of logic is required in order to write or list things of importance. It is essential to provide the examiner with a mature impression of yourself and to show that you are capable of communicating your thoughts sensibly. Sentence construction, grammar, spelling, punctuation, etc., tell the examiner so much about the candidate. Students who are dyslexic should inform the college examination office with a letter from their doctor and this should be sent to City and Guilds well before the examination date.

Calculation questions need students to lay out their work in an orderly fashion, starting off with an introductory word to begin a formula which is often derived from information in the question. Words like: **since, therefore, and, hence, also,** etc., help the logic of the calculation as it proceeds towards the answer. Numerous examples can be found in this book. It is essential not to use the calculator until all the facts from the question are written down. An answer should be preferably underlined (not in red ink) and the appropriate unit (if any) stated. There is no need to produce the answer again on another line. Clear layout is very important. An equals sign should be placed on the same line as the quotient line (the line which divides the numerator from the denominator) and all workings should be shown.

Drawing questions need to be thought out. Start by using a pencil. If a circuit diagram is required, use a ruler and drawing aids to show circles for instruments. Remember that lines can cross each other but when joining, a dot is used to show this. Labelling is vitally important and preferably should be neatly printed with an arrow to identify the part or component it illustrates. Sketches, too, need to be labelled and it should be mentioned here that most of these diagrams can easily be drawn from the simple outline of a cube in some realistic size. The sketch needs to look like the actual object and must incorporate all the important features.

Appendix 2

Laboratory Work

With the present Course 236 Part II national examination results being under 63%, there is a real need to make students not only more communicative in their written work but also more knowledgeable in their understanding of basic principles. The highest failure rate is in the second written paper with a third of the examination candidates failing this component. One comment made by the City and Guilds examiners concerned the question of impedance and resistance. It said:

> 'A very popular question but very few correct answers. Part (a), about impedance, was clearly not understood by most candidates. A large number stated that it was the total resistance of a circuit.'

This is typical of an area which lends itself admirably to a laboratory exercise which allows students to investigate the nature of circuits comprising resistance, inductance and capacitance. It is the intention of the author, in this Appendix, to explain to students how laboratory experiments should be written up and in so doing, involve them in more awareness of their technical studies so that they are able to communicate their findings on paper.

Writing Reports

There is no set standard for writing laboratory reports but some laboratory books do include useful hints on this procedure. It is felt by the author that laboratory work should be done by students after they have acquired the necessary information and notes on the topic being investigated, at which stage students will be in a better position to understand the experiment and relate it to some principle or theory.

The layout recognized by many college lecturers is as follows:
1. Title
2. Object
3. Apparatus
4. Circuit Diagram
5. Method
6. Results
7. Conclusions

The title, object, apparatus and circuit diagram are often provided for students to copy. The rest will require students' ability to complete. Briefly, the **title** of the experiment should be placed in the middle of the laboratory book at the top, with the date on the right-hand side. The object and other sub-headings should be written in the margin on the left-hand side and all these headings, including the title, should be underlined with a ruler (not in red ink).

The **object** should state exactly what the experiment is hoping to achieve or hoping to investigate. The **apparatus** should be a list of all the equipment used in the experiment and should clearly name the equipment as well as its electrical rating (if any). The **circuit diagram** should be clearly drawn with a ruler and other aids showing all connection points for instruments and the supply. The circuit should of course be fully labelled.

The *method* should be written around the instructions which students will follow on the exercise sheet. It should be written in the past tense using no personal pronouns such as 'I', 'we' and 'our', etc. It should follow the logic of carrying out the experiment, from the commencement of wiring components up to the eventual finish where all readings are noted and results obtained. The **results** should be set out in a tabulated form (preferably boxed in) and they should portray both observed values and calculated values, both sets being clearly illustrated and labelled. At this stage in the report, graphs and charts can be drawn and they also need to be properly labelled with appropriate title and scale data. It is much better to do this work in pencil and show neat printing. Graphs of two variables should not be formed by lines joining dot to dot; they need to be formed from lines of best fit and students should consult their lecturers where doubt exists. Students should get used to the terms 'proportional to' and 'inversely proportional to' as well as 'saturation point' and 'linear', etc. The most difficult part of writing up an experiment is the **conclusion**. Firstly, look back at the object and state if this was achieved or verified. Secondly, make reference to any observed results, for example, was there any change in instrument readings? Was any heat produced or was any magnetic effect noticed? Thirdly, what did your results produce? Did, for example, the power factor improve as a result of more capacitance and did the supply current value fall? Did the graph show that the motor had a constant speed up to full load? Did the graph of the carbon filament lamp show that it had a negative temperature coefficient of resistance? The conclusions are very important; they may relate to some formulae, principle or law or they may relate to a requirement from a statutory document, such as the 1988 Electricity Supply Regulations, or a non-statutory document, such as the 1981 IEE Wiring Regulations. The following examples are typical experiments to enable students to actively participate in this area of the electrical syllabus.

Experiment 1 (Example Report)

Resistance and Capacitance in Series

Object

To investigate the current and voltage relationship in an a.c. series circuit containing resistance and capacitance.

Apparatus

1 × 0/5 A ammeter
1 × 0/150 V voltmeter
1 × 0/115Ω/2.8 A variable resistor
1 × 100 μF capacitor

Circuit Diagram

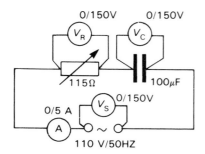

Figure 218 *RC series circuit*

Method

The circuit was connected as shown above. The variable resistor was set to its maximum full-in position. After checking circuit connections the supply was switched on. The resistor was then adjusted so that the ammeter read 1.0 A. The voltmeter was used to measure the potential difference across (i) supply terminals, (ii) variable resistor and (iii) capacitor. All values of voltage were recorded. The above procedure was repeated for current values of 2.0 A and 3 A.

The observed and calculated results for the three conditions are shown below.

Results

Observed				
V_S	V_R	V_C	I_S	C
113	109	30	1	100
113	96	62	2	100
113	66	91	3	100

Calculated				
X	R	Z	PF	ϕ
30	109	113	0.96	15°
31	48	57	0.84	33°
30	22	38	0.58	55°

Where: V_S is the supply voltage
V_R is the p.d. across the resistor
V_C is the p.d. across the capacitor
I_S is the supply current
X_C is the reactance of the capacitor
R is the resistance of the resistor
Z is the impedance of the circuit
PF is the power factor of the circuit
ϕ is the phase angle between supply voltage and current.

Note The above calculations are made from course notes.

Phasor diagrams representing the three conditions are shown overleaf.

Conclusions

The relationship between current and voltage in this circuit can be summarized as follows: As the current is allowed to increase in the circuit by adjustment of the variable resistor, the less resistance, the smaller becomes the potential difference across the resistor.

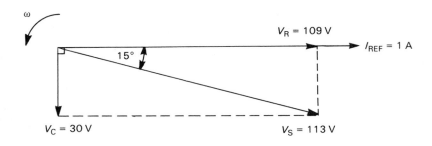

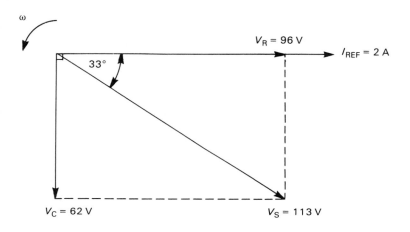

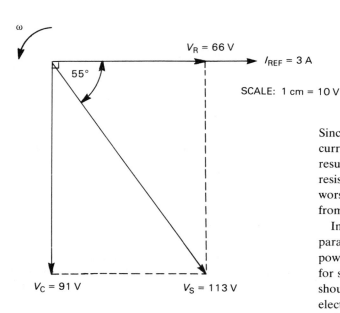

Figure 219 *Phasor diagrams for a.c. circuit*

Since the capacitor has a fixed value, the increase of current increases its potential difference. The above results and phasor diagrams show that, with less resistance in the circuit, the power factor becomes worse. This is seen by the phase angle increasing from 15° to 55°.

In practice, capacitors are normally connected in parallel with inductive loads to improve lagging power factor conditions. It should also be noted that for safety reasons, power factor correction capacitors should be fitted with a means for discharging their electrical energy – such as discharge resistors (see IEE Regs., Reg. 461–4).

Experiment 2 (Example Report)

Resistance and Inductance in Series

Object

To investigate the current and voltage relationship in an a.c. series circuit containing resistance and inductance.

Apparatus

1 × 0/5 A ammeter
1 × 0/150 V voltmeter
1 × 0/115Ω/2.8 A variable resistor
1 × 0.09 H/3.26Ω inductor

Circuit Diagram

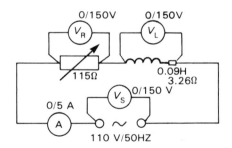

Figure 220 *RL series circuit*

Method

The circuit was connected as shown above. The variable resistor was set to its maximum full-in position. After checking circuit connections the supply was switched on. The resistor was then adjusted so that the ammeter read 1.5 A. The voltmeter was used to measure the potential difference across (i) supply terminals, (ii) variable resistor and (iii) inductor. All values of voltage were recorded. The above procedure was repeated for a current value of 2.5 A.

The observed and calculated results for both conditions are shown below.

Results

Observed					
V_S	V_R	V_L	I_S	R_I	L
113	98.5	45	1.5	3.26	0.09
113	76	72.5	2.5	3.26	0.09

Calculated							
V_L	V_{LX}	V_{LR}	X_L	R	Z	PF	ϕ
43	42.4	4.9	28.3	65.7	74.5	0.92	23°
71.5	71	8.15	28.3	30.4	44	0.76	40°

Where: V_S is the supply voltage
V_R is the p.d. across the resistor
V_L is the p.d. across the inductor
V_{LX} is the p.d. across inductor's reactance
V_{LR} is the p.d. across the inductor's resistance
I_S is the supply current
R_I is the resistance in the inductor
L is the inductance of the inductor
X_L is the reactance of the inductor
R is the resistance of the variable resistor
Z is the total impedance of the circuit
PF is the power factor of the circuit
ϕ is the phase angle between supply voltage and current.

Note The above calculated results are made from coursework notes. Phasor diagrams representing the conditions are shown in Figure 221.

Conclusions

The relationship between current and voltage in this circuit can be summarized as follows: The main observations to make from this experiment are that

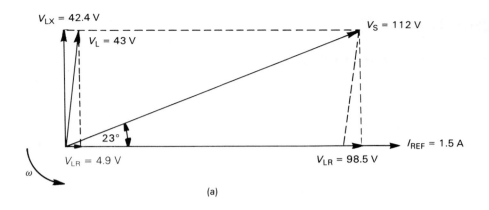

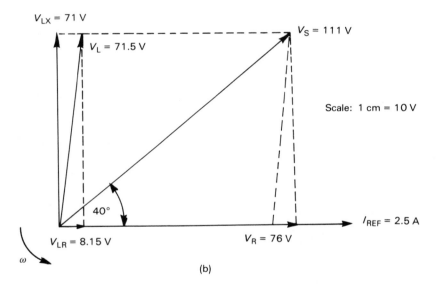

Figure 221 *Phasor diagrams for a.c. circuit comprising resistance and inductance in series*

as the current increases in the circuit from 1.5 A to 2.5 A, the circuit resistance becomes less and the potential difference across the variable resistor falls. Since the inductor has a fixed value, the increase of current increases the potential difference across the inductor. The second phasor diagram shows that the phase angle between supply current and supply voltage has increased to 40°. In simple terms, this means that the power factor of the circuit has become worse, falling from 0.92 lagging down to 0.76 lagging.

It should be pointed out that since the inductor comprises resistance and reactance, the p.d. across the inductor (V_L) measures the voltage across the inductor's impedance. The values of V_{LX} and V_{LR} relate to the inductor's internal p.ds and from both phasor diagrams the coil is seen to be highly inductive. Further investigations will show that if the variable resistance was completely cut out of circuit, then the inductor would have a much worse power factor, falling to 0.114 lagging. The supply current would lag behind the supply voltage by a phase angle of 83.4°.

Experiment 3 (Example Report)

Power Factor Improvement

Object

To investigate the power factor and current of a single-phase a.c. circuit supplying an inductive load when a variable shunt capacitor bank is added to the circuit.

Apparatus

1 × 0/2 A ammeter
1 × 0/150 V voltmeter
1 × 0/20Ω/5 A rheostat
1 × 100 μF capacitor bank
1 × 120 V/2.5 A wattmeter
1 × 7Ω/0.2 H inductive coil

Circuit Diagram

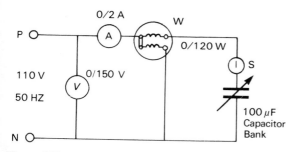

Figure 222

Method

The circuit was connected as shown above with the capacitor bank switch open. After checking all connections a supply was given to the inductive load and instrument readings noted. The capacitor bank switch was then closed and capacitance introduced into the circuit in steps of 5 μF to a maximum of 80 μF. For each addition of capacitance, instrument readings were taken and recorded.

The observed and calculated results are shown below. A graph was plotted to show the relationship between current, power factor and capacitance.

Results

	Observed			Calculated	
C	P	V	I	PF	φ
0	84	112	1.68	0.446	63.5°
5	84	112	1.52	0.493	60.5°
10	84	112	1.40	0.535	57.6°
15	84	112	1.24	0.605	52.7°
20	84	112	1.12	0.669	47.9°
25	84	112	0.99	0.757	40.7°
30	84	112	0.90	0.833	33.6°
35	84	112	0.81	0.926	22.2°
40	85	112	0.76	0.998	3.0°
45	85	112	0.76	0.998	3.0°
50	85	112	0.78	0.973	13.3°
55	85	112	0.85	0.893	26.7°
60	85	112	0.95	0.799	36.9°
65	85	112	1.05	0.722	43.7°
70	85	112	1.18	0.643	49.9°
75	85	112	1.33	0.570	55.2°
80	85	112	1.48	0.513	59.1°

Where: C   is the capacitance introduced into the circuit
P   is the power consumed in the circuit
V   is the supply voltage
I   is the current taken by the circuit
PF   is the power factor ($\cos \phi$)
φ   is the phase angle between circuit current and voltage

Note The above calculated results are made from coursework notes.

Conclusions

The observations made from this experiment show that while the wattmeter readings kept reasonably constant the supply ammeter readings decreased to 0.76 A. The results table shows that this was for a capacitance of between 40 μF and 45 μF. From the graph Figure 223 (and by calculation), a 43 μF capacitor is required to bring the power factor of the circuit to unity conditions. This would cause minimum current to flow in the circuit and remove the phase angle difference between current and voltage. The capacitor required for unity power factor would need to be rated 170 VA.

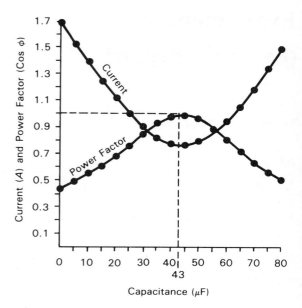

Figure 223

Experiment 4 (Example Report)

D.C. Shunt Motor

Object

To investigate the speed–torque and efficiency characteristics of a d.c. shunt motor.

Apparatus

1 × 0/2.5 A ammeter
1 × 0/250 mA ammeter
1 × 0/150 V voltmeter
1 × Jay-Jay motor testing set
1 × stroboscope

Circuit Diagram

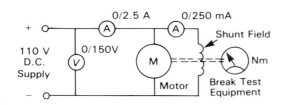

Figure 224

Method

The motor's circuit was connected as shown above. After checking all connections a supply was given to the motor. Adjustment was made to the motor's brake so that its torque indicated 0.04 Nm. The machine was then allowed to run for ten minutes to reach its correct operating temperature. The speed, ammeter and voltmeter readings were noted as the motor's brake was increased in steps of 0.04 Nm to stalling point. The following results were obtained and from collected data, graphs plotted of speed and efficiency.

Results

Observed				Calculated		
T	N	V	I	P_I	P_O	Eff
0.04	37.17	110	0.25	27.5	9.33	0.34
0.08	36.83	110	0.35	38.5	18.51	0.48
0.12	36.33	110	0.50	55.0	27.38	0.50
0.16	35.83	110	0.56	61.6	36.01	0.58
0.20	35.17	110	0.52	68.2	44.18	0.65
0.24	35.00	110	0.58	74.8	52.76	0.71
0.28	34.67	110	0.77	84.7	60.97	0.72
0.32	33.83	110	0.85	93.5	68.01	0.73
0.36	33.33	110	0.94	103.4	75.38	0.73
0.40	33.32	110	1.02	112.2	83.71	0.75
0.44	33.32	110	1.10	121.0	92.08	0.76
0.48	31.67	110	1.20	132.0	95.48	0.72
0.52	31.67	110	1.28	140.8	103.44	0.73
0.56	31.50	110	1.40	154.0	110.22	0.72
0.60	30.00	110	1.49	163.9	113.07	0.69
0.64	30.00	110	1.60	176.0	120.61	0.68
0.68	29.83	110	1.70	187.0	127.44	0.68
0.72	28.33	110	1.85	203.5	128.15	0.63
0.76	0	110	O/L	O/L	O/L	

Where:
T is the motor's torque in newton-meters
N is the motor's speed in rev/s
V is the supply voltage
I is the current taken by the circuit
P_I is the motor's input power
P_O is the motor's output power
Eff is the motor's per unit efficiency

Conclusions

It will be seen from the observed results that as more torque is placed on the motor's brake its speed decreases, stalling at 0.76 Nm. While the supply voltage remained constant, the current in the circuit increased from 0.25 A to 1.85 A. The calculated results show the input power and output power both increasing as the motor becomes fully loaded. The efficiency also increases reaching a maximum condition at 0.44 Nm – beyond this loading it starts to fall.

The speed–torque graph shows that the motor operates at approximately constant speed, falling only 4 rev/s at maximum efficiency.

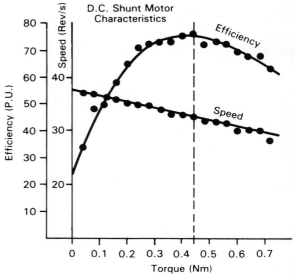

Figure 225

Experiment 5 (Example Report)

Efficiency of Electric Kettle

Object

To investigate the efficiency of an electric kettle and to account for any heat losses.

Apparatus

1 × 1500 W/240 V copper electric kettle
1 × 0/15 A ammeter
1 × 0/250 V voltmeter
1 × 500 ml measuring beaker
1 × stop clock
1 × Celsius thermometer

Circuit Diagram

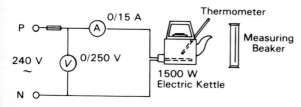

Figure 226

Method

The kettle was filled with 2 litres of cold tap water and then connected to the supply as shown above. The initial temperature of the water was taken and found to be 20°C. The main supply was then switched on and the stop clock started simultaneously. The ammeter and voltmeter readings were recorded and the kettle switched off when the water temperature reached boiling point. The experiment was repeated for 1 litre of water making sure that the kettle and element were properly cooled before refilling. The following results were obtained.

Results

	Observed					Calculated		
	V	I	m	$\delta\theta$	t	W(kJ)	Q(kJ)	Eff
Exp. 1	240	6.2	2	80	500	744	672	0.90
Exp. 2	240	6.2	1	80	300	446.4	336	0.75

Where:
- V is the supply voltage
- I is the current taken by the circuit
- m is the mass of water
- $\delta\theta$ is the temperature rise in the water
- t is the time in seconds
- W is the kettle's input energy
- Q is the kettle's output heat energy
- Eff is the kettle's per unit efficiency

Note:

$W = V \times I \times t$ joules
$Q = m \times c \times \delta\theta$ joules
c is the specific heat capacity of water (4200 J/kg K) and 1 litre of water is equivalent to 1 kg.

Conclusions

It will be seen from the observed results that the kettle element was operating at 1488 W instead of 1500 W and that the kettle boiled in 5 minutes with one litre of water compared with 8.33 minutes with two litres of water. The experiment showed that the more water contained in the kettle the more efficient it is. Heat is mostly lost by conduction through the kettle's metal enclosure and the more full it is the longer it takes to heat up since the element is using the water as a conducting medium. Other losses occur through convection and evaporation as the kettle nears and finally reaches boiling point.

Further investigations could be made with different capacities of water and the calculations of efficiency will improve if consideration is given to the heat required for the copper container.*

---

*An example of this is given on page 112–113 of EIT Vol. 2

Appendix 3

Part I Assignments

The following is a typical City & Guilds assignment which electrical installation students have to do for the Part I Certificate.

The assignment is to be completed in approximately 5 hours of college time and students should provide themselves with the necessary drawing materials, drawing aids, textbooks, catalogues and a copy of the IEE Wiring Regulations for electrical installations.

Site Huts

Specification

The attached drawing (Figure 227) is of two flat roofed huts on a small site. One hut is the office, the other for use by the site personnel. The 240 V supply to the huts is from the supply unit adjacent to the office. Each hut is provided with an S.P. and N miniature circuit breaker consumer unit.

The rating of the luminaires and appliances are as follows:

interior lights	100 W
outside lights	60 W all-insulated bulkhead luminaires
oil-filled radiators	2 kW
tubular heater	1.5 kW
tea urn	3.0 kW

The mounting heights of the accessories above floor level are as follows:

lighting switches	1400 mm
switch socket outlets	1000 mm
20-A D.P. switches	500 mm

The installation is to be carried out using p.v.c.-insulated and sheathed cables.

Section A – Pre-planning

1. The following items of plant are sent direct from another site. State the checks required to ensure that they are in a safe and usable condition.
    (a) Bending machine incorporating a pipe vice.
    (b) A pair of 8-tread wooden steps.

Section B – Measurement and setting out

2. Prepare an 'overlay' to show the routes of the cables for the lighting installations.
   (Note: outside lights are via a joint box.)
3. By measurement from the overlay, estimate the length of cable required to carry out the lighting installation.

Section C – Installation planning

4. Make a sketch showing how the cable is to be run to the site personnel hut.

Appendix 3

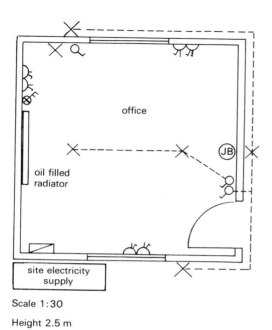
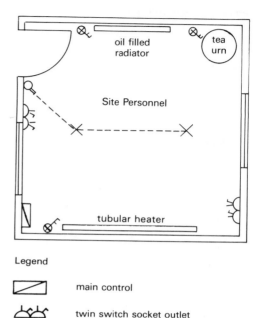

Scale 1:30
Height 2.5 m

Figure 227 *Site huts*

Legend

	main control
	twin switch socket outlet
	20 A.D.P. switch with pilot light
	tungsten luminaire

5. Draw the interior connections of the joint box for the outside lights, identifying all cables and conductors.

Section D – Communications

6. The pins holding the blades of a new hacksaw fall out and are lost in the process of changing the blade. Write a letter of complaint to the manufacturer, requesting a replacement hacksaw.

Section E – Testing and inspection

7. State the purpose of a polarity test on a new installation.
8. Describe with the aid of sketches, how to carry out the polarity test on the lighting installation of the office.

Section F – Problem solving

9. Neglecting thermostatic action, calculate the total energy consumed in a week if:

The tea urn is 'on' for 4 hours a day over 5 days.
The tubular heater is 'on' for 24 hours a day over 7 days.
The oil-filled radiators are 'on' for 10 hours a day over 5 days.
The hut lighting is 'on' for 6 hours a day over 5 days.
The outside lighting is 'on' for 3 hours a day over 5 days.

Solution

The answers below are extracted from the assignment carried out by a Part I, second-year, day-release student working for an electrical contracting firm.

Section A – Pre-planning

1. (a) In checking a bending machine incorporating a pipe vice, I would firstly open the machine out and make sure it stands firmly, without any movement in the hinge at the top. I would then make sure the arm used to pull down and bend the conduit is directly in line with the former in which the conduit is placed, to ensure true bends. There must also be a 'stop' present to secure the conduit when being bent. The 'former' on the arm which bends the conduit must be well rounded to stop the conduit from being marked or dented.

   On the vice, I would make sure that all the teeth are present, not worn and firmly fixed to the jaws of the vice. The thread of the vice used to alter the size required between the jaws must be free from dirt or swarf and be well greased.

   After these checks are carried out an overall check is made to make sure all nuts and bolts, joints, etc., are secure and there are no fractures in the metalwork of the structure.

   (b) With a pair of wooden steps, I would firstly open them out on a flat surface and make sure they stand level and firm. I would check the hinges at the top of the steps and make sure they are not worn and are securely screwed to the woodwork. The rope used to distance the legs of the steps must be of equal length and securely knotted at either end of the legs. The rope must not be worn or frayed. If chain is used, the lengths must be equal and fastened securely and all links of the chain must be sound.

   After these checks have been carried out, an overall check is made to make sure that all the woodwork is in order with no splinters or cracks, checking each of the eight steps individually making sure that every step is secure.

Section B – Measurement and setting out

2. See Figure 228.
3. Length of cable for lighting in the office hut is as follows:
   3-core pvc/pvc/cpc 10 m (approx).
   2-core pvc/pvc/cpc 10 m (approx).
   Length of cable for lighting in site personnel hut is as follows:
   2-core pvc/pvc/cpc 10 m (approx).

Section C – Installation planning

4. See Figure 229.
5. See Figure 230.

Section D – Communication

6. Letter of complaint.

A. J. Swann (U.K.) Electrical Contractors Ltd
80 Highbury Corner, Bedford MK44 666
(0099) 66655

M.D. Ltd
Engineering Tool Manufacturers
Channel House
123–125 Slip End Warf
Bute Town
Milton Keynes MK45 4BB

22nd November 1989

Dear Sir,
I am writing to inform you of the 'Swift' G400 hacksaw which was recently purchased (see copy AC Order CC/102/0098 enclosed). After one week of using the hacksaw the pins holding the blade have fallen out and are now lost. I personally believe there is a manufacturing fault in that the pins are too loose. Without these pins the tool is obviously redundant.

In exercising good customer relations, I would be grateful if you could forward a replacement hacksaw to our site as soon as possible.

Yours faithfully,   *L. B. O'Brian*

Mr L. B. O'Brian
(Contracts Manager)

Appendix 3

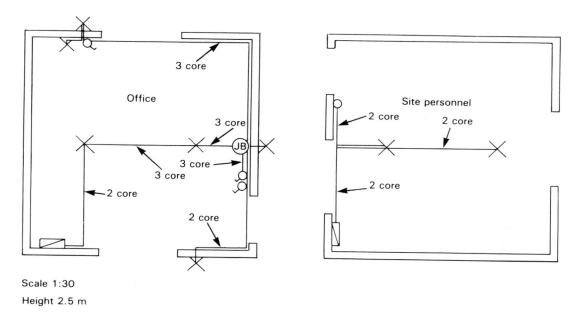

Figure 228 *Cable routes for lighting circuits*

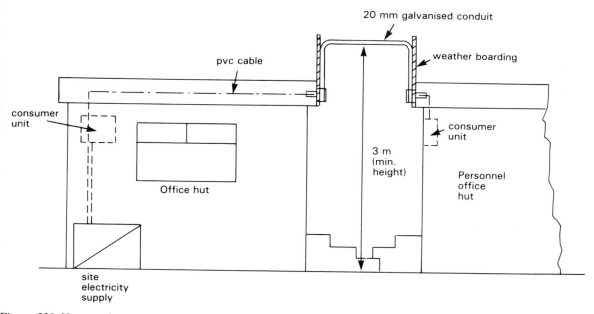

Figure 229 *Hut supplies*

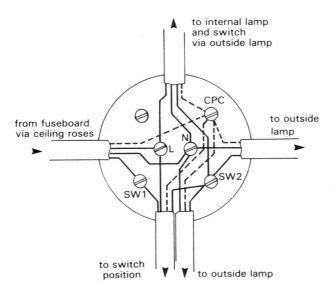

Figure 230

Section E – Testing and inspection

7. A polarity test must be carried out on a new installation to verify that all fuses, miniature circuit breakers and single pole control devices, such as switches, are connected in the phase conductor only. The centre-contact of an Edison-type screw lampholder should be connected to the phase conductor and the outer metal threaded part should be connected to the neutral conductor.

   In any electrical installation it is important to identify and correctly label the polarity of the circuit conductors in conjunction with the requirements specified in Section 524 and Tables 52A and 52B of the IEE Wiring Regulations.

8. Figure 231 shows the circuit diagram of how a polarity test can be made using a proprietary digital ohmmeter. The test is made before the main supply is energized. In the distribution board, a link is inserted between the phase conductor and cpc and with the lamp removed and local switch in the 'on' position, the ohmmeter probes are attached to the ceiling rose (Test A). If the circuit polarity is correct then the meter will indicate a very low ohmic value ('dead-short'). The test can also be done at the switch (Test B). When identified, the bare cpc should be sleeved with green/yellow sleeving and if a black insulated conductor is used as a switch wire it has to be identified with a red disc or red cable sleeve.

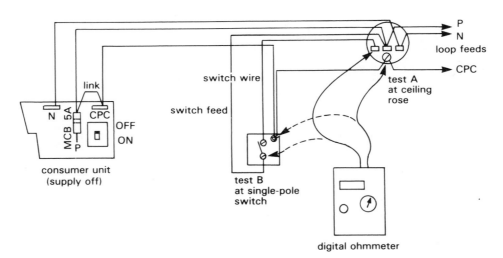

Figure 231

9. **Section F – Problem solving**

The formula for finding Energy (W) is
$W = $ Power (P) $\times$ time (t) joules

Tea urn
$W = Pt = 3\,000 \times 4 \times 3\,600 \times 5 = 216\,00$ MJ
Tubular heater
$W = Pt = 1\,500 \times 24 \times 3\,600 \times 7 = 907.20$ MJ
Oil-filled radiators
$W = Pt = 4\,000 \times 10 \times 3\,600 \times 5 = 720.00$ MJ
Interior hut lighting
$W = Pt = 100 \times 5 \times 6 \times 3\,600 \times 5 = 54.00$ MJ
Exterior hut lighting
$W = Pt = 60 \times 3 \times 3 \times 3\,600 \times 5 = 9.72$ MJ
Total energy used = 1906. 92 MJ
Since 1 kWh = 3.6 MJ then total energy over one week = 529.70 kWh

Appendix 4 Possible course study route at a technical college

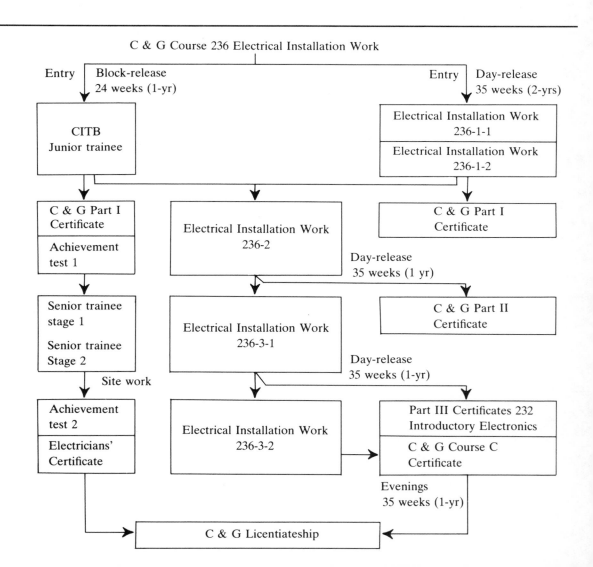

Appendix 5

Common Deviations in Electrical Installations

Reg. No.	Deviations
13–1	Good workmanship and proper materials shall be used. No compliance.
314–1(ii)	Neutral conductor not connected in the same sequence as the phase conductor.
314–4	Borrowed neutrals.
412–14(ii)	RCD not rated at 30 mA for equipment used out of doors fed from a 13-A socket outlet.
413–2	Omission of a main bond to water, gas, other pipe, risers or structure.
413–8/10	Exposed conductive parts of an installation not earthed.
434–2	Quality of the fuse or MCB not suitable for the prospective short circuit current.
461–3	Warning notice to indicate more than one means of isolation on switches and equipment.
467–15	Omission of main switch or circuit breaker.
467–17	Appliance or luminaire not in readily accessible position.
471–35	Omission of bathroom or shower room bonding of pipes and the supplementary bonding of infra-red heater.
471–38	Omission of Home Office skirt at a bath or shower room lampholder.
471–47	Requirements for socket outlet (32 A or less) for use outdoors.
511–1	Equipment not in compliance with British Standards.
512–1/2	Equipment not suitable for a.c. voltage and current.
512–6	Equipment to be suitable for the mode and situation – check to be made on I.P. number.
514–1	Identification labels for switchgear and controlgear to avoid confusion.
514–2	No identification of protective devices, e.g. m.c.bs.
514–3	No provision of charts, diagrams or tables and use of BS 3939 installation symbols.
514–4	Voltages greater than 250 V to be indicated, e.g. 415 V.
514–5	Omission of the periodic inspection label, complete with dates.
514–5(a)	Omission of the quarterly RCD test notice.
514–7	Safety label omitted at a bonding clamp/electrode/structure bond.
514–8	Omission of a label to indicate that a socket is suitable for outside use.
522–8	Voltage drop not considered in the installation design.
523–8	Installation not suitable for damp situations – steel screws used outside, etc.
523–10	Aluminium and brass/copper conductors used without protection.
523–15/16	Ingress of water or dust in an enclosure owing to blank plates missing.

523–19	No mechanical protection for cables, e.g. channel at skirting level.		terminations. Entry holes made too large – especially at ceiling roses.
523–20	Holes in joists for cables not being 50 mm from the top or bottom of joists.	528–1	Omission of making good holes in walls and ceilings.
		529–1	Insufficient clipping of cables – usually in a roof void.
523–20 (a–d)	Cables buried in walls not being horizontal or vertical.	529–2	Insufficient support for conduits and trunking.
523–21	Protection from abrasion, e.g. use of grommets, bushing strip or brass bushes.	543–1	Adiabatic equation or Table 54F not considered in the selection of protective conductors.
523–23	Omission of cable covers or marker tapes for underground cables.	547–2	Inadequate c.s.a. of main bonding conductors.
524–3	Conductor termination identification, e.g. red/yellow/blue.	547–3	Main bond connection not as near as practicable to the point of entry.
525–3/4/5/6	Mixed category circuits.	547–5	Inadequate c.s.a. of supplementary bonding conductors.
525–7	Omission of barriers in dual boxes for 13 A/T.V. socket outlets.		
527–1	Poor joints and/or omissions of cord grip devices.	552–4	Omission of solenoid type starter for a motor where automatic starting might cause danger.
527–2	Undersize terminations or cables terminated without suitable lugs.	614–1	Failure to issue Completion and Inspection certificates.
527–5	Inadequate enclosure of cable		